Erosion-Corrosion:
An Introduction to Flow Induced Macro-Cell Corrosion

Written By

Masanobu Matsumura

Professor Emeritus, Hiroshima University
237-101 Fukumoto, Saijo-cho, Higashi-Hiroshima 739-0031
Japan

eBooks End User License Agreement

CONTENTS

Foreword   *i*

Preface   *ii*

CHAPTERS

    1.   **Historical Review of Research on Erosion and Erosion-Corrosion**   3

    2.   **Pure Erosion Processes-Cavitation and Solid Particle Impact Erosion**   18

    3.   **Combined Erosion and Corrosion**   49

    4.   **Theory of Electrochemical Corrosion**   70

    5.   **Erosion-Corrosion Testing Methodology**   94

    6.   **Case Study of Erosion-Corrosion in the Field**   118

    7.   **Prevention of Erosion-Corrosion in the Field**   137

        Index   157

FOREWORD

Flow influenced materials degradation is encountered in many fields of our daily and technical experience, starting from rivers carving themselves into rock formations and finishing with severe wall thinning in metal pipes and pumps of industrial plants. Understanding the flow and materials related basic steps of initiation and continuation of such degradation processes has been subject of many research efforts in the last decades. This was often prompted by severe failures resulting in high financial losses and sometimes even fatalities. However, even today after all this work done so far in this field there is still a debate about how to describe the interactions between materials and flowing media in a proper and predictive way. It starts with the definition of terms used to classify different reasons for flow influenced materials attack: cavitation, erosion, corrosion, flow accelerated corrosion, etc. and combinations thereof. Worldwide, national and international standards give different definitions based on different views of the main influencing parameters.

This book starts with the intention not to add to this Babylonian confusion but to explain step by step the different phenomena and contribute to the understanding rather than verifying definitions of terms. This is one of the big merits of this book. The reader is firstly taken on an interesting travel through the historical approaches of research in cavitation, erosion and erosion corrosion which for many decades was significantly influenced by the author's excellent and innovative work in this field. The somewhat narrative style in which the chapters are written keep the reader fascinated and eager to learn more about the experimental evidences for later conclusions. Many of the intelligent experimental devices and protocols for studying flow influenced materials degradation, presented and explained throughout the book, have been developed in the author's research group over the years. I am proud to say at this place that they also inspired me some 20 years ago when I had the pleasure to meet Prof. Masanobu Matsumura in Germany. The exciting discussions we had at that time prompted us to design and produce jet impingement devices to study erosion corrosion phenomena electrochemically with jet impinged microelectrodes resulting in our "freak event" energy density theory for erosion corrosion initiation.

The book is an excellent summary of the great and path breaking research achievements of the author and his research group in the complex field of erosion corrosion and can only be warmly recommended for everyone who wants to understand the effect of influencing parameters in cavitation and erosion corrosion, alone and in combination with each other. It is an outstanding source of information for those who want to learn about erosion corrosion testing methods. Case studies underline and backup the technical importance of flow influenced corrosion in its various localized and wall thinning appearance and its threat for failures in industrial plants. The book fills a gap that has been open for too long time.

Günter Schmitt
Ceo Ifinkor
Institut fuer Instandhaltung und
Korrosionsschutztechnik gGmbH
Germany

PREFACE

Erosion-corrosion is a generic name for the degradation phenomenon that occurs in chemical plants, in which metallic materials are exposed to various flowing liquids. For example, erosion-corrosion occurs within heat-transfer pipes made of copper-based alloys that are used in seawater heat exchangers. Moreover, it also takes place within both casings of seawater pumps made of gray cast iron and carbon steel pipes that transport pure water at high temperature and pressure, *etc.* Erosion-corrosion was the likely cause of a serious accident in the United States in 1986 that killed four workers at Surry nuclear power plant located in the state of Virginia.

A half century ago, in 1962, the author was a freshman of the master course of Tokyo Institute of Technology. His supervisor was Professor Yoshitada Suezawa who belonged to the department of chemical engineering. He instructed the students that the mechanical engineers in the field might be disposed to classify the erosion-corrosion in corrosion, but the chemical engineers classify it into erosion, so the boundary region of erosion and corrosion must be the underdevelopment region to be elucidated. He also recommended the student to boldly begin with the opposite side of the region, that is, the cavitation erosion.

For nearly five decades, the author has expanded his research field from cavitation erosion to the whole area in the boundary region of erosion and corrosion, and made a recent notable discovery regarding the nature of erosion-corrosion. Contrary to the popular belief, erosion-corrosion in nature does not result from cooperation between erosion and corrosion; instead, it manifests from pure electrochemical corrosion, which should be classified as localized corrosion or macro-cell corrosion. This fact was proven by the author using both experimental and theoretical evidence. Experimental data showed no cooperation product between erosion and corrosion even when pure erosion and pure corrosion overlapped. Thus, it was concluded that erosion-corrosion is a pure corrosion process, rather than a cooperative process. A theoretical analysis using an Evans diagram indicated that erosion-corrosion is macro-cell corrosion that results from at least two different anodic polarization behaviors of the metal which are originated from differences in the flow conditions.

This book is sparing much space on erosion-corrosion testing methods and presents case studies conducted on accidents that have occurred at actual chemical plants. These topics were the focus of the research and, simultaneously, the means for reaching the conclusions: erosion-corrosion damage can only be studied in the laboratory with the appropriate testing equipment; and, erosion-corrosion data from accidents in the field are still needed to study the mechanisms responsible for erosion-corrosion. Testing methods cannot be developed without complete clarification of the mechanisms responsible for erosion-corrosion damage.

For many years, the author belonged to the chemical engineering department of Hiroshima University. One of the purposes of the chemical engineering is to support the chemical industries in the field. Therefore, the end-result of this research will not be the elucidation of the mechanism by which erosion-corrosion degrades structural materials in a chemical facility. Rather, identification of the mechanisms of erosion-corrosion will enhance the ability to predict both the locations of accidents in chemical plants and the degree of damage that might result. In addition, identification of the erosion-corrosion mechanisms will be useful for both the design and material selection of hydraulic machines and chemical plants, such as for seawater pumps and piping in nuclear power plants, and for maintenance and safety management.

Masanobu Matsumura
Higashi-Hiroshima, Hiroshima
E-mail: mmatsu@rapid.ocn.ne.jp

CHAPTER 1

Historical Review of Research on Erosion and Erosion-Corrosion

Abstract: By referring the history of the research in each field, cavitation erosion and erosion-corrosion were compared. The start of the research on the cavitation erosion was the generation mechanism of the cavitation impulsive pressure, which brought about the damage to metallic materials. From both sides of impulsive pressure generation mechanism description by theory and measurement of impulsive pressure by the experiment, the generation of cavitation erosion damage has been recognized to be a pure physical or pure mechanical process. As to erosion-corrosion, corrosion must possibly play the leading part in the damage. However, as to the agency of erosion component and its role has not been clarified yet. The shear force of the fluid flow is the most probable agency of erosion. Nevertheless, it is rather irrational to attribute the various forms of erosion-corrosion solely to the shear force of fluid flow over the metal surface.

Keywords: Erosion, erosion-corrosion, cavitation erosion, impulsive pressure, corrosion, shear force, horseshoe corrosion, impingement attack, inlet tube corrosion, deposit attack, turbulence corrosion.

1. TERMINOLOGY FOR CAVITATION EROSION

According to the famous equation of Bernoulli (1754), the static pressure drops when the flow velocity of fluid rises. Liquid is used to boil when the static pressure drops to the level of vapor pressure at the temperature, but the static pressure can be lowered, depending on conditions, without boiling lower than the vapor pressure, and still further to the negative pressure (tension). Then, the liquid may suddenly be ruptured into many small bubbles (cavities) which are filled with the vapor of the liquid. This is cavitation.

In the field of technology, the harmful influence of cavitation was recognized in the year 1895 when the high-speed destroyer of United Kingdom, which was equipped for the first time with the steam turbine, could not demonstrate expected performance: the propeller rotated at high speed but not enough thrust could be obtained since a large amount of bubbles was generated. The test running was forced continued, and the propeller was totally consumed in a day and night. The cause of the depletion of propeller might be the large impulsive pressure which was generated when the cavitation bubbles collapsed.

Lord Rayleigh (1917) first estimated theoretically the intensity of the impulsive pressure [1]. He assumed a spherical bubble which would reduce its volume keeping its shape spherical as shown in Fig. (**1-1**). A small portion of liquid at the bubble wall surface must have a small amount of kinetic energy as it approaches the center of bubble, which is integrated for whole volume of the liquid to obtain the total kinetic energy of the whole of the liquid. This may be converted into the compression energy when the bubble stops reducing its size. The pressure of thousands atm was estimated to occur when a bubble reduced to one twentieth of the original diameter.

Fig. (1-1). Process of cavitation bubble collapsing [1].

Around 1960, a new theory asserted that the bubble would not keep its shape spherical in the decay but would destroy the form by some unstable factors producing a liquid jet which approaches the inside of the bubble. Based on the theory, Plesset and Chapman [2] estimated the velocity of the liquid jet, which is being originated at the collapse of the bubble which is attached on the solid wall, to be as high as 128 m sec^{-1} (Fig. (**1-2**)). In addition to this, the developing process of the liquid jet was observed with a high speed motion picture camera.

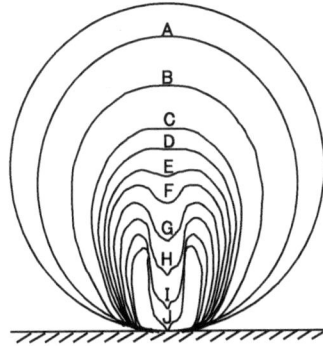

Fig. (1-2). Unsymmetrical decay of bubble accompanied with liquid jet formation [2].

The classic theory of Lord Rayleigh was, however, not removed by the new theory, but both theories are equally accepted in the recent days, that is, cavitation bubbles are believed to collapse in both ways in every case. Important is that the intensity of the impulsive pressure must be widely distributed in any way it may be originated: it is easily imagined that the intensity of impulsive pressure would depend on the diameter as well as the collapsing rate of the bubble. Further, the level of the impulsive force, which the solid wall surface receives, must be widely distributed depending on the distance between the collapsing bubble and the wall surface, and the direction of the liquid jet as well.

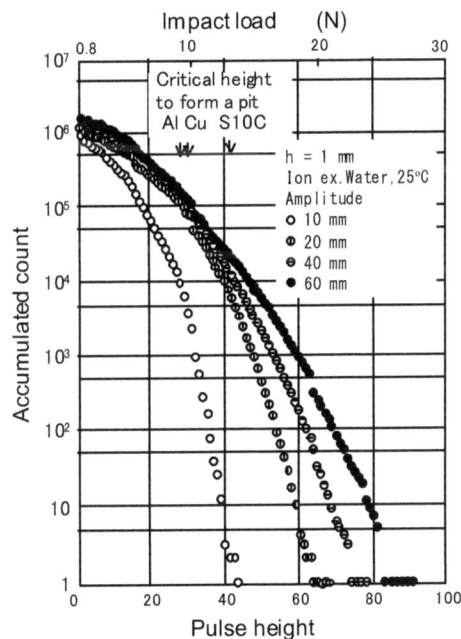

Fig. (1-3). Distribution profiles of cavitation impulsive pressure determined in the vibratory test apparatus with the piezoelectric device: measurements are being given in the unit of load, N, as the surface area pressed is unable to be determined; S10C, the carbon steel containing 0. 10 % C [3].

In the latter half of the 1980's, by the development of personal computer and piezoelectric device, it became possible to measure the distribution of the impulsive pressure of cavitation. Fig. (**1-3**) is an example of typical impulsive pressure distribution measured by Iwai *et al.* in the vibratory testing equipment which is to be described in the next section [3]. The generation frequency of impulsive pressure in the *y*-axis of the diagram is given in a logarithmic scale. Though the impulsive pressure with low intensity arises innumerably, the generation frequency of impulsive pressure of such high intensity to cause plastic deformation in the surface of metallic material with the attack of only one time is remarkably low.

Throughout over half century, the pursuit of the generation mechanism of cavitation impulsive pressure and the measurement of the intensity of the pressure, as described above, have been conducted mainly by physicists and mechanical engineers. They, in imagining, wish to confirm that the origin of damage to metallic materials be a pure mechanical process, that is, erosion. According to the terminology of ASTM (American Society of Testing and Material) (ANSI/ASTM G 40-77), "erosion" was defined as *progressive loss of original material form a solid surface due to mechanical interaction between that surface and a fluid, a multicomponent fluid, or impinging liquid or solid particles.*

On the other hand, chemists seemingly believed, from the beginning of the first recognition of cavitation, that there is an involvement of corrosion in the process of cavitation damage: they must possibly claim that without any doubt corrosion is concerned in the degradation of the metallic propeller in the seawater which bears the highest corrosiveness. They did not change the view even after the cavitation testing equipments suitable for the testing of various materials, which shall be described in the next section, were developed, and the cavitation damage was brought about to such materials as glasses, rock crystals, Bakelite and gold that do not corrode. In the International Standard, ISO 8044^{1999} with the title of Corrosion of metals and alloys — Basic terms and definitions, any separated term "erosion" or "cavitation" are not yet listed, but the term "cavitation corrosion" is defined as *process involving conjoint corrosion and cavitation*, and term "erosion corrosion" is defined as *process involving conjoint corrosion and erosion*. Thus, it seems that in the field of chemistry, erosion is not a separate terminology as it is equally ranked with corrosion, but it is considered to be some modification of corrosion.

2. DEVELOPMENT OF CAVITATION EROSION TESTING EQUIPMENT

In the early days of cavitation research, the glass U-tube filled with liquid was used for studying the generation mechanism. The water tunnel was utilized for the observation of incipient cavitation around propellers and torpedoes.

In the 1960's, the vibratory cavitation testing equipment, as shown in Fig. (**1-4(a)**), suitable for the testing of various materials, was invented. Though there was the anxiety that the cavitation generation mechanism of the equipment might differ from that of actual hydraulic machines, it was adopted for the first time as the standard testing method of ASTM (ANSI/ASTM G32-77) because it was small in volume, little in consuming power as well as in quantities of test liquid. According to the standard method, a disc shaped test specimen 16 mm in diameter is allowed to vibrate immersed in the test liquid by 10-30 mm with amplitude of 50 μm and a frequency of 20 kHz.

The intensity of cavitation which is excited under the conditions was comparatively high, so that some measurable damage occurred in several hours even to the specimen of tough material. Thus, this testing equipment is convenient as an accelerated testing method (rapid test method) as it is possible to finish the test in a short period of time. The measurement of the intensity distribution for the impulsive pressure shown in Fig. (**1-3**) was conducted by using the cavitation equipment of this type. It can be easily imagined that the cavities do not repeat the process of generation and disappearance in each cycle of the test specimen vibration.

In the 1970's, the stationary specimen vibratory cavitation test facility as shown in Fig. (**1-4(b)**) was invented, in which the specimen was held just under the vibrating probe (dummy specimen) and the cavities generated at the probe surface by its vibration were moved to the specimen surface below to collapse there and result the impulsive pressure. Though the intensity of cavitation attack on the specimen surface was, of course, lower than that of on the vibrating specimen, it possessed the remarkable advantage that the large acceleration was avoided which would be otherwise inevitably imposed on the vibratory specimen repeatedly with the high frequency. It had, however, another disadvantage in the accuracy of the measurement which might come from the rise in the temperature of the specimen and the probe as well as the liquid in between.

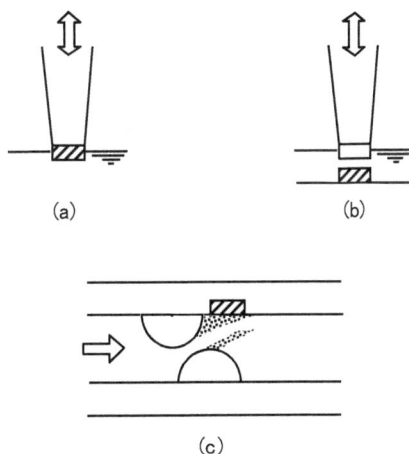

Fig. (1-4). Options of cavitation test facilities for material evaluation: (a) vibratory specimen facility; (b) stationary specimen vibratory facility; (c) water tunnel.

In 1973, the water tunnel suitable for conducting cavitation erosion test on various materials was developed by Louis [4]. In the rectangular flow channel, two weirs of semi-cylindrical shape were arranged so that the flow of liquid containing cavities struck the specimen surface which was installed behind the weir downstream (Fig. **1-4(c)**). The intensity of the impulsive pressure of cavitation excited on the specimen was high enough for the test duration being shorten, and accordingly for itself being accepted as an acceleration test equipment which could simulate the cavitation damage in the actual hydraulic machines.

In the year 1979, this author [5] and his collaborators were engaged in the improvement of the stationary specimen vibratory cavitation test facility in order to elucidate the effects of tensile stress on cavitation damage: it could be easily assumed that the parts of practical machines where cavitation erosion occurred must be under some stress such as centrifugal force due to rotation, residual stress after working or stress by its own weight. At that time, various results had already been reported concerning the effects of stress on erosion, which did not necessarily reach the same conclusion. This was partially because different materials and apparatuses were used for the experiments. But the greatest cause for the diversity of the experimental result was probably because the effects of stress on the damage were so little that they fell within the errors of measurement of the amount of damage. Hence, they started their investigation by improving the prototype of stationary specimen facility in order to raise the accuracy of the measurement.

The items improved were (1) circulation of the testing liquid by pumping it through a bore to the gap between the stationary specimen and the vibrating prove (made of type 304 stainless steel) in order avoid the temperature rise in the horn which must otherwise cause the thermal expanding of it, which inevitably results in the change of the distance between the specimen and the probe, and accordingly in a poor reproducibility of the test results, (2) making use of a self-aligning ball bearing in order to hold the probe and the specimen exactly parallel, as shown in Fig. (**1-5**). The specimen was installed on a table equipped with the piston through ball bearing and jacks. These jacks were first loosened so that the table could freely change its direction from the action of the ball bearing. Then the table was lifted by adjusting the screw in order to press the stationary specimen lightly to the probe.

Thus, the two surfaces were set parallel by contacting each other directly. After the table was fixed to the piston by three jacks, the piston was lowered by the vertical adjusting screw to set the gap between the specimen and the probe.

Fig. (**1-6**) shows an example of experimental results of sufficient reproducibility obtained after the improvement described above. The amount of scatter in the relationship between the weight loss of specimen, W (mg), and the testing time, t (min), for four specimen of aluminum was within ± 3 %.

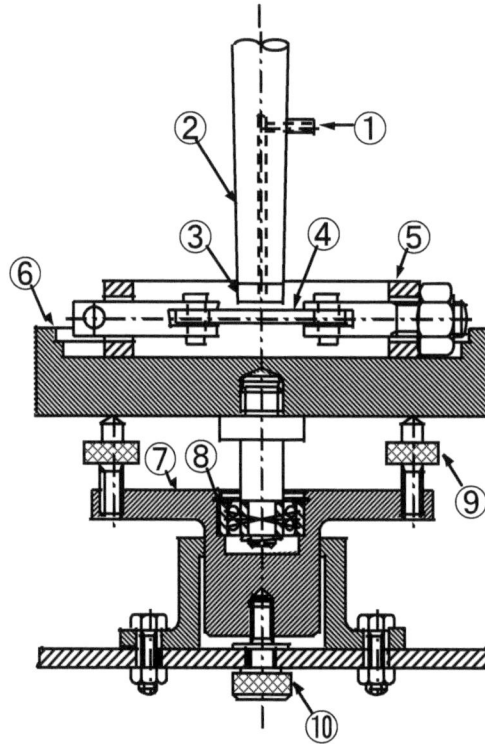

Fig. (1-5). Drawing of stationary specimen vibratory test facility: 1. Test liquid, 2. Horn, 3. Vibrating probe, 4. Stationary specimen, 5. Loading frame, 6. Table, 7. Piston, 8. Self-aligning ball bearing, 9. Jack, 10. Vertical adjusting screws [5].

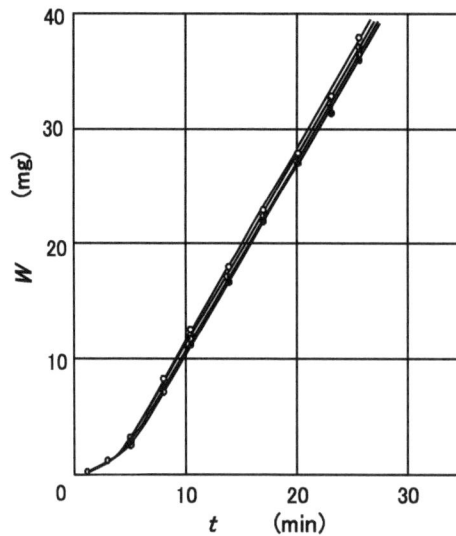

Fig. (1-6). Weight loss versus testing time curves of aluminum specimens tested under the same conditions: frequency, 19. 9 kHz; amplitude, 28 μm; separation, 0. 4 mm; flow rate of test liquid, 0. 5 L min^{-1} [5].

Tensile stress was applied to the dumbbell-shaped stationary specimen with the simple and compact device shown in Fig. (**1-7**). The results of tests on aluminum specimen under the application of tensile stress are shown in Fig. (**1-8**). The curve drawn with a broken line shows the relationship between the testing time and the cumulative weight loss of one specimen. Other curves with tensile stress applied to the specimen were obtained by several specimens: for instance, six were used to obtain the curve of 2. 2 Kg mm^{-2}. Comparing the curve of broken line with the solid line of 3. 1 Kg mm^{-2}, you can see the exact converse effects of tensile stress on cavitation damage: the favorable effect of suppressing cavitation damage, and the unfavorable effect of accelerating the damage depending totally on with or without the loading-unloading repetition. In the successive experiments on iron, mild steel and cast iron, the repetition of loading-unloading was, of course, avoided.

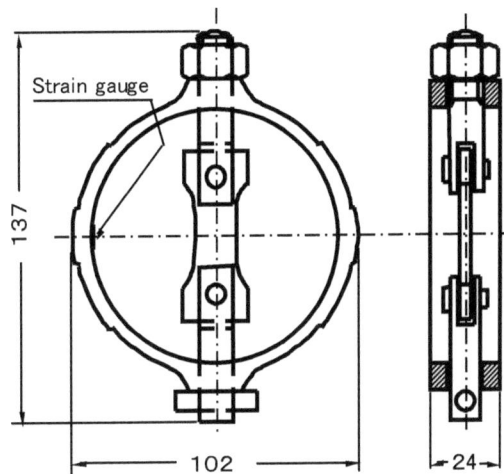

Fig. (1-7). Loading frame [5].

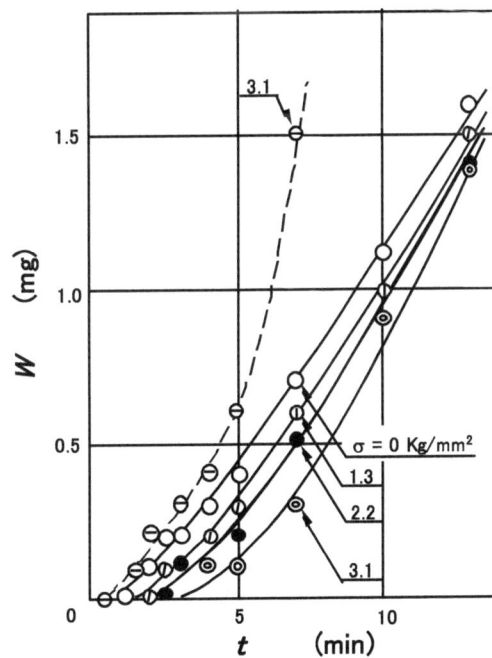

Fig. (1-8). Cavitation damage depressing effect of tensile stress applied to aluminum specimen [5].

The *W vs. t* curve was differentiated to obtain the weight loss per unit time, that is, the cavitation erosion rate, *R* (mg min^{-1}). The *R vs. t* curves for aluminum specimens are shown in Fig. (**1-9**). All the curves bear clear incubation periods. For the ductile materials of aluminum, iron and mild steel, the incubation period was held longer as higher level of stress was applied: the incubation period of aluminum with the tensile stress of 3. 1 Kg mm^{-2}, for instance, was extended five times as long as that of the same material without the stress. On the contrary, the incubation period for cast iron specimen, which was a typical brittle material, was reduced by the application of tensile stress.

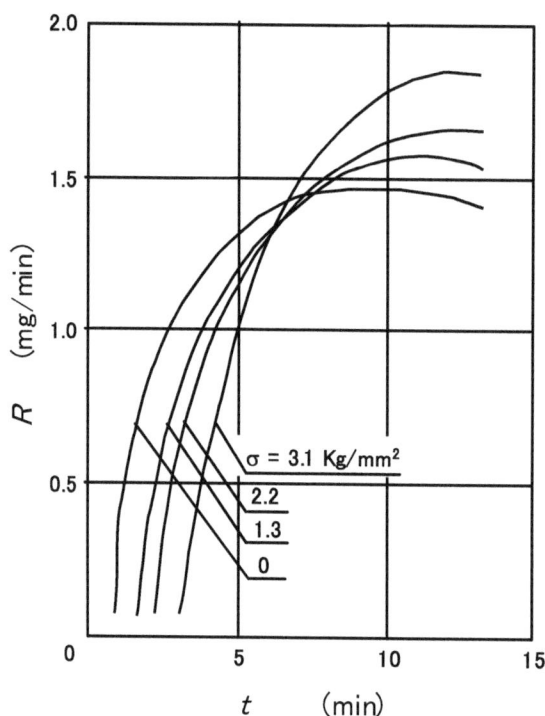

Fig. (1-9). Increase in incubation period with increasing tensile stress applied on aluminum specimen [5].

Another noticeable effect of tensile stress on cavitation erosion was that the erosion rate *R* increased more rapidly and reached at a higher peak as a higher level of tensile stress was applied to the specimen. The same tendencies were observed on the behavior of iron, mild steel as well as cast iron. So, it was concluded that tensile stress applied to ductile metals generally prolonged the incubation periods. In contrast to this favorable effect, it also brings about the unfavorable effect to accelerate the cavitation damage in the period of weight loss, regardless of the properties of metals.

3. EROSION-CORROSION OCCURRING ON VARIOUS METALS

3.1. Localized Corrosion on Seawater Condenser Tubes of Copper Alloy

All the thermal and nuclear power stations in Japan are located on the seacoast. The reason for this is the utilization of seawater, which is inexhaustible and free, to cool the steam which is exhausted from the power generating turbines and condensed into pure water.

Seawater is so corrosive that the condenser tubes of the heat exchangers have to be made of copper-base alloys, which are known to be resistant to seawater at rest. In flowing seawater, however, localized corrosion sometimes occurs, so various technologies have been developed to protect the condenser tubes from corrosion, and new types of copper-base alloys with anti-flowing-seawater corrosion properties have been developed. Recently, tubes of titanium, which are almost completely resistant to seawater corrosion, have been used for the steam condensers in power plants.

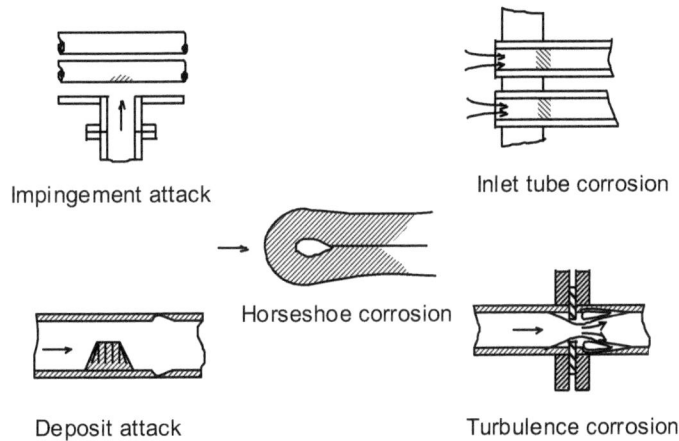

Fig. (1-10). Types of localized corrosion that arose on copper alloy tube surface in flowing seawater.

Fig. (**1-10**) shows the types of localized corrosion that occur on copper alloy tubes in flowing seawater. A detailed description of each type follows.

> **impingement attack** — this occurs at the outside surface of heat exchanger tubes that are arranged at the entrance of the shell side fluid, where it collides with the tubes at a right angle. This type of corrosion damage occurs even when the impinging fluid does not contain any suspended solid particles or cavitation bubbles.

> **deposit attack** — this occurs at the floor wall surface as well as the ceiling surface, both downstream from such deposits as barnacles, but not at the ceiling surface immediate overhead where the flow cross section narrows most and consequently the shear stress reaches maximum.

> **horseshoe corrosion** — this occurs as tiny dents at the entrance of copper alloy condenser tubes. The size is small, less than 1 mm in width, and the length depends on the flow conditions. The appearance is of a short hoof-print of horse, which is the derivation of the name.

> **inlet tube corrosion** — this is a type of localized corrosion which occurs in the entrance of the tube at one or two times the tube diameter downstream from the inlet end of the tube, but not at the outlet end.

> **turbulence corrosion** — localized corrosion damage is often found at locations downstream from the orifice or the flow nozzle. This is assumed to be due to the flow turbulence there, which is therefore the derivation of the name.

The types of localized corrosion mentioned above occur to a restricted extent on the metal surface, and their names were mainly derived from the shape of the corrosion damage or the location where it occurred, except for turbulence corrosion, which may be derived from the presumed occurrence mechanism.

3.2. Erosion·Corrosion Mechanism Proposed by Syrett

In 1976, Syrett [6], who occupied an important position in the National Association of Corrosion Engineers (NACE), introduced the generation-mechanism-based terminology, "erosion-corrosion", in order to integrate the visual terminology in the field, as shown in Fig. (**1-10**).

The dot between the words "erosion" and "corrosion" means "and", so one might misunderstand the terminology to include erosion, which means the separation of metal fragments from the surface due to physical or mechanical forces as described in the preceding section. Nevertheless, Syrett clearly denied the existence of an erosion component, and illustrated its generation process as shown in Fig. (**1-11**). The corrosion rate increases gradually with increasing flow velocity or with the shear stress of flowing fluid. At a certain velocity, which Syrett called "breakaway velocity", the protective oxide layer over the metal surface is separated by the shear stress of flowing fluid, and the base metal surface is exposed to the environmental liquid. This is connected with the formation of a macro-cell of corrosion, where the base metal surface is the macro-anode and the oxide layer the macro-cathode, which further increases the corrosion rate. The macro-cell of corrosion disappears when the oxide layer is completely separated, at the same time the corrosion rate reaches a plateau.

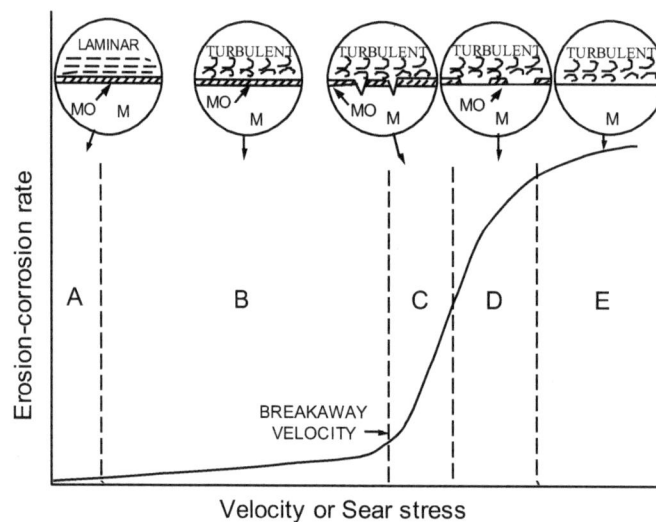

Fig. (1-11). Illustration of the generation process of erosion-corrosion proposed by Syrett [6].

3.3. Differential Flow-Velocity Corrosion on Cast Iron

Among the various types of pumps used for different purposes, seawater pumps are mostly of a conventional type and are apt to suffer corrosion damage. Adoption of higher-grade materials is very effective in reducing corrosion damage, but corrosion problems cannot be completely overcome. It is, therefore, desirable from an economic point of view to use common materials, such as plain cast iron, in combination with skillful anti-corrosion engineering.

In 1986, Kitajima *et al.* [7], who was a corrosion engineer at a pump manufacturing company in Japan, had the following experience. The corrosion rate of cast iron in seawater at rest is usually about 0. 2 mm y^{-1} and increases with elevating flow velocity. On the casing surface of a cast iron pump, however, corrosion damage was mild at the periphery of the spiral part where the fluid velocity is high, and was conversely so severe near the shaft hole, where the velocity is low, that the maximum depth of graphitization (typical corrosion damage to plain cast iron) exceeded 2 mm after operation for 4 600 h.

In order to investigate this phenomenon, which opposes the concept of erosion-corrosion, Kitajima *et al.* cut the casing and cover of the pump down into small fractions, then put leading wires on each piece and bonded them together with epoxy resin to rebuild the casing, the cover, and the pump. While pumping seawater, corrosion current was measured using a zero-shunt ammeter between a certain fraction (shown by dotted lines in Fig. (**1-12**)) and the rest of the fractions. The result of the current measurements clearly substantiated that the low velocity zone near the shaft hole remained anodic. Thus, the formation of a large-scale corrosion cell was recognized. They called this "differential flow-velocity corrosion".

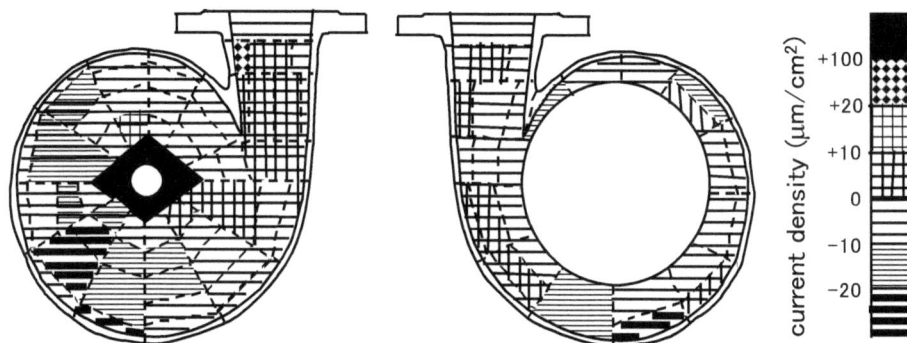

Fig. (1-12). Distribution of corrosion-current over the pump-casing of cast iron [7].

3.4. Wall Thinning of Carbon Steel Pipes in Deoxygenated Pure Water

It is not unusual for a certain sort of wall thinning to occur at the rate of over 1 mm y^{-1} on the inside wall surface of carbon steel pipes that carry pure water of high electric resistivity (0. 1-1×10^6 Ω cm) in the elevated temperature range of 100-200 °C. The notable feature of this wall thinning is the extent over which it is generated: it spreads rather uniformly over the whole circumference of the pipe at a distance of several times the pipe diameter along the pipe axis. For example, in the year 2000 it was found in Japan that, in carbon steel piping system at a waste heat recovery boiler, after seven years of operation, 3 of 18 locations downstream from the orifice meter suffered this sort of wall thinning. The following features of this wall thinning have been reported:

1. The thinning rate ranges from 0. 5 to 1 mm y^{-1}. It may, of course, sometimes be lower, and can sometimes reach as high as 4 or even 9 mm y^{-1}.

2. A number of parameters influence the rate of wall thinning, among which the most important is the temperature of the environmental pure water, the details of which will be described later. Other influencing parameters include water chemistry issues: pH and dissolved oxygen content (DO) in the liquid; flow conditions such as fluid velocity, mass transfer rate and the geometry of the flow or the pipe fittings; and, material specifications such as the content of chromium or molybdenum in carbon steels.

3. Wall thinning arises quite accidentally; it sometimes occurs and sometimes does not, even when the parameters mentioned above are identical. A typical example is as follows. Three boilers of an identical set-up were in parallel operation under identical conditions. Wall thinning took place in only one of them. No trace of wall thinning was, however, recognized at the corresponding positions on the pipelines of the other two boilers.

With respect to temperature effects, the experiences of engineers in the field and the results of laboratory tests are in close agreement. The bell-shaped behavior of the wall-thinning rate of carbon steel in pure water, with a narrow peak at 140-150 °C as shown in Fig. (**1-13**), is typical. The sharpness of the characteristic peak dulls with decreasing flow rate, until it diminishes at a flow rate of 302 kg h^{-1}, indicating clearly that the effect of fluid temperature on the rate is influenced by the factor of fluid flow. Actually, almost all the hot water pipe rupture accidents have occurred in that temperature range. The reliability of Fig. (**1-13**) is thus very high.

Nevertheless, there are three minor questions regarding this figure. The first relates to the bell-shaped behavior of the temperature effect on the rate. Several attempts have been made, without success, to explain it by assuming that the carbon steel surface is covered with a thick layer of magnetite (Fe_3O_4) and the mass transport of Fe^{2+} species away from the oxide/solution interface is the controlling process in determining the rate. However, no good reason has been given as to why the magnetite solubility turns out to decrease with increasing temperature at ca. 140 °C, and furthermore, why this behavior depends on the flow rate.

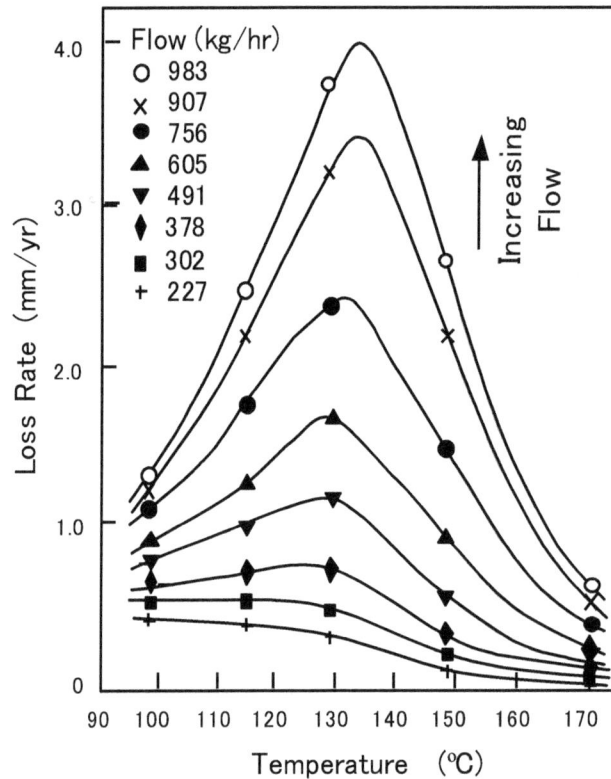

Fig. (1-13). Characteristic temperature dependency of wall-thinning rate [8].

The second question is as follows: the loss rate on the ordinate of the diagram exceeds 1 mm y^{-1} for those flow rates that are higher than 605 kg h^{-1} for nearly the entire range of temperature. Consequently, according to the diagram, carbon steel pipes could not be used for quite a wide range of those conditions, since the wall thinning rate of 1 mm y^{-1} is the upper limit for an industrial material to be used in the field. This aspect of the diagram is not in agreement with the situation in the field, in which carbon steels are widely and satisfactorily used in pure water environments at elevated temperatures without any serious problems except for the accidental occurrence of wall thinning.

The third question is why the flow rate, but not flow velocity, was chosen as the parameter. The effects of fluid flow on the temperature dependency of the corrosion rate should be expressed in terms of velocity or shear stress, just as Syrett did.

As to the pH effect on wall thinning, it has been found that an increase in the pH of the water above 9. 2 causes a decrease, by more than one order of magnitude, in the erosion-corrosion rate of carbon steels, Fig. **(1-14)**. This diagram promises complete release from erosion-corrosion in water with pH values higher than 9. 5. As a matter of fact, a boiler manufacturing company in Japan has recommended this countermeasure to its customers. The response, however, was not positive. There must be some inevitable reasons for which the pH of boiler water cannot be raised.

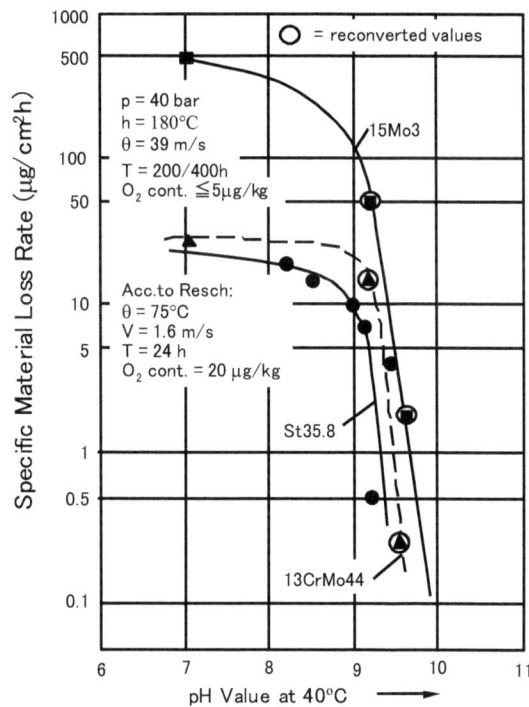

Fig. (1-14). pH dependency of erosion-corrosion rate for different steels [9].

3.5. Computer Programs Estimating the Rate of Flow-Accelerated Corrosion (FAC)

For the last several decades, the most commonly accepted mechanism for wall thinning of carbon steels in flowing high-temperature deoxygenated pure water has been the physical dissolution of the magnetite film enhanced by mass transport of soluble Fe (II) species, where (II) is the oxidation number, or Fe^{2+} ions away from the surface as given by Eq. (1-1) in Fig. (**1-15**). The reasons why this physical mechanistic model was deduced seem to be as follows:

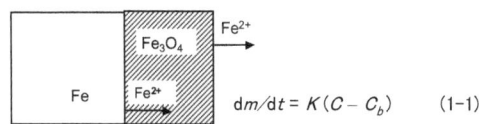

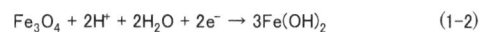

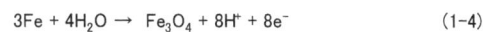

$$dm/dt = K(C - C_b) \qquad (1\text{-}1)$$

$$Fe_3O_4 + 2H^+ + 2H_2O + 2e^- \rightarrow 3Fe(OH)_2 \qquad (1\text{-}2)$$

$$Fe \rightarrow Fe^{2+} + 2e^- \qquad (1\text{-}3)$$

$$3Fe + 4H_2O \rightarrow Fe_3O_4 + 8H^+ + 8e^- \qquad (1\text{-}4)$$

$$CR = F_1(T)\ F_2(AC)\ F_3(MT)\ F_4(O_2)\ F_5(pH)\ F_6(G)\ F_7(\alpha) \qquad (1\text{-}5)$$

Fig. (1-15). EPRI model illustrating the formation of flow-accelerated corrosion (FAC) [10].

1.   It was believed that the electrochemical dissolution of metals was not possible in pure water with a high electric resistivity.

2.   In pure water that is flowing in nuclear power plants, a phenomenon called CLUD occurs, in which fine particles of radioactive iron oxide or iron hydroxide increase in number with the lapse of operating time.

A major flaw in the physical dissolution model, that is, the FAC model, was pointed out by the members of the Electric Power Research Institute (EPRI): according to Eq. (1-2) in Fig. (**1-15**), the dissolution of magnetite to form Fe (II) species is a cathodic reduction reaction, and hence cannot proceed unless it is balanced by an anodic oxidation reaction [10]. Further, when the only anodic reaction available in this regard, that is Eq. (1-3) in Fig. (**1-15**), is added to the model, it turns out to be a complete electrochemical corrosion model, which is contradicted by the first term above. Thus, when EPRI was asked by the Nuclear Regulatory Commission (NRC) to plan countermeasures to prevent the recurrence of the Surry accident, there was no reliable model that could be applied to the wall thinning of carbon steel pipes.

EPRI therefore, had to recommend the most conservative, and reliable, method: early detection of the reduction in pipe wall thickness through the nondestructive inspection and prophylactic replacement of the components. They intended to promote the recurrence prevention plan by reducing the burden on the engineers who had to inspect more than 30 000 spots in a power plant, which lie under a thick heat-insulating layer over the entire outside surface of the pipe. It was necessary to separate the suspected area that needs to be precisely and frequently inspected, from the area where the inspection interval may be prolonged. In order to achieve this purpose, excellent prediction techniques were necessary to estimate the wall-thinning rate at different spots from plant operating parameters.

They began with the separation-of-variable type equation, Eq. (1-5) in Fig. (**1-15**), where corrosion rate, CR, is independently correlated with each parameter, such as the water temperature (T), the content of alloy element in the steel (AC), mass transfer (MT) factors, *etc.* All the available data from twenty nuclear plants in the United States, as well as laboratory data from Electricité de France (EDF) in France and the Central Electricity Generating Board (CEGB) in England, were used to develop the model empirically by an iterative procedure until an optimum correlation was obtained. Thus, the first version in the computer-aided FAC program series was completed.

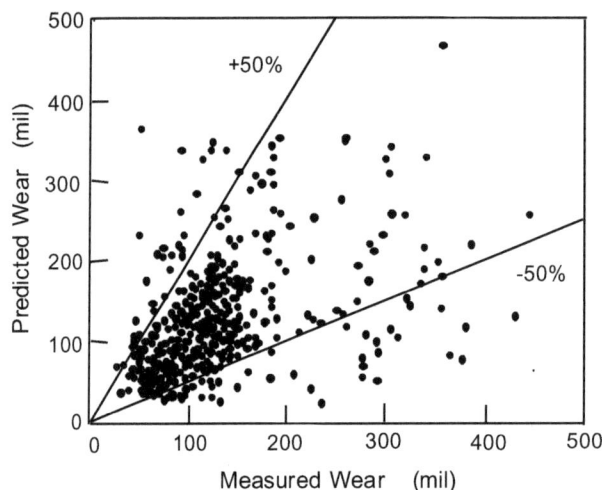

Fig. (1-16). Performance of FAC computer program in predicting wear damage [11].

Fig. (**1-16**) shows the performance of one of these programs against the plant data. EPRI was satisfied with the agreement for the moment, considering the uncertainties in pipe operating conditions in the field. However, they soon decided to improve the accuracy of the prediction through an endless use of wall thinning data obtained from the inspection of power plants worldwide, including those in Japan.

Concerning the relationship between the terms erosion-corrosion and FAC, there is an interesting description in a footnote of the book which had the title of "Flow-Accelerated Corrosion in Power Plants", which was written by the related members of EPRI and EDF [11], and published in 1996: *In the United*

States, flow-accelerated corrosion is commonly but incorrectly known as erosion-corrosion. The members of EPRI and EDF might have the idea that these were generated from a common origin, in spite of the major differences between erosion-corrosion and FAC in the pipe materials and the temperature ranges in which they occur, and particularly in the corrosiveness of seawater and pure water.

4. STANDARDS AND THE RESEARCH

The standardization is indispensable for the sound development of the technology and the industry accordingly. Or, it would be no exaggeration to say that the standardization itself is the development of technology. In the research on the degradation phenomenon of industrial material, the standardization of terminology and testing methodology is the fruit of research: without the clarification of real mechanism of the material degradation phenomenon, it would not be rightly termed, and any testing method for materials cannot be properly specified either. In reverse to this, the standardization of terminology and the specification of testing method may be instituted in advance in order to stimulate the development of the research and elucidation of mechanism. For the standards instituted with this purpose, reviewing these standards with the development of the research is natural and necessary work, and in fact, standards are revised widely and frequently as a matter of course. Judging from the viewpoint of standardized terminology and testing methodology, there may be considerable difference in the progress situation of the research between erosion and erosion-corrosion: as to the terminology in the field of erosion-corrosion, there are some terms which must be seemingly named arbitrarily while the mechanism is not yet proven, and some confusion accordingly. In order to avoid any more confusion, no more new term should be proposed here. So, therefore, the terms which are necessary in this book were restrictedly chosen and defined as follows, obeying mainly the ASTM standard.

cavitation — the formation and collapse, within a liquid, of cavities or bubbles that contain vapor or gas or both.

erosion — progressive loss of original material from a solid surface due to mechanical interaction between that surface and fluid, a multi-component fluid, or impinging liquid or solid particles.

cavitation erosion — progressive loss of original material from a solid surface due to cavitation impulsive pressure.

erosion-corrosion — should not be yet here defined, as it is the main theme of this book.

It was quite recently that the new work item with the title of "Guidelines for the selection of methods for erosion-corrosion testing in flowing liquids" was, at last, proposed to the ISO/TC156 Corrosion of Metals and Alloys. As you can see, this is apparently the latter sort of standardization mentioned above. Probably eternal time will be necessary for the completion of a new standard of testing methodology for erosion-corrosion, because every testing method, though it is not limited to the standardization, must be made to satisfy two requirements: the reliability and the rapidness. The former means that the measurements can be obtained with small scattering, and the latter that the measurement can be finished in a short time. The scattering of the measurements may possibly be reduced by technical ingenuity and contrivance as it was demonstrated in the case of the stationary specimen vibratory cavitation testing equipment that was described in Section 2 of this Chapter. The shortening of testing time, however, cannot be achieved without greatly accelerating the damage rate. Then, the rate-controlling parameter must inevitably be altered. At the very moment of the alternation, the dilemma begins: how closely does the testing equipment simulate the damage in the field? This is the severe destiny of a rapid test methodology to receive the pursuit on this problem.

There may be two ways to prove whether the rapid testing equipment simulates the actual damage of the machines and plants in the field: one is to take it as a black box, and to compare the merit and demerit ranking of various materials obtained through the testing equipment with the performance of those materials in the field; the example of this will be shown in Chapter 5. Another way is to compare the damage mechanism in the test equipment with that in the actual machines of the field, the example of which will be given in the next chapter.

REFERENCES

[1]   Rayleigh L. On the Pressure development in a liquid during the collapse of a spherical cavity. Philosop Mag J Sci, Sixth Series 1917; 34: 94-98

[2]   Plesset MS, Chapman RB. Report No. 85-09; Office of Naval Research 1970.

[3]   Iwai Y, Okada T, Tanaka S. A study of cavitation bubble collapse pressure and erosion Part 2: Estimation of erosion from the distribution of bubble collapse pressure. Wear 1989; 133 (2): 233-43

[4]   Louis H. Erosive Zerstörungen durch Strömungskavitation. Doktor-Ingenieur genehmige Dissertation, Technische Universität Hannover, 1973.

[5]   Matsumura M, Okumoto S, Saga Y. Effects of tensile stress on cavitation erosion. Werkst. u. Korr. 1979; 30: 492-8.

[6]   Syrett BC. Erosion-corrosion of copper-nickel alloys in sea water and other aqueous environments —A literature review. Corrosion 1976; 32: 242-52.

[7]   Kitashima N, Ichikawa K, Kinoshita K, Miyasaka M. Corrosion in seawater pumps and its prevention. Boshoku-Gijutsu (presently Zairyo-to-Kankyo) 1986; 35: 633-41.

[8]   Bignold GJ, Garbett K, Garnsey R, Woolsey IS. Tackling erosion-corrosion in nuclear steam generating plant. Nucl. Eng. Int. 1981: 37-41.

[9]   Heitmann HG, Kastner W. Erosionscorrosion in Wasser-Dampfkreisläufen. VGB Kraftwerkstech. 1982; 62: 211-9.

[10]   Cohen P. Ed. The ASME handbook on water technology for thermal power systems. New York: ASME 1989; pp. 947-67.

[11]   Chexal B, Horowitz J, Jones R, Dooley B, Wood C, Bouchacourt M, Remy F, Nordmann F, Paul PS. Flow-Accelerated corrosion in power plants. EPRI report TR-106611. Pleasant Hill, CA: Electric Power Research Institute (EPRI) 1996.

CHAPTER 2

Pure Erosion Processes-Cavitation and Solid Particle Impact Erosion

Abstract: Damage depth rather than the weight loss of specimen was adopted for representing the extent of cavitation damage to metallic materials. As a result, linear relationship was obtained between the testing duration and the extent of the damage, that is, the damage depth. Furthermore, from this linear relationship a characteristic index was obtained which represented the resistance of material to cavitation attack. In order to represent the extent of damage during the incubation period, where neither weight loss nor damage depth were observed, another parameter named surface increment percentage was introduced. As a result, another characteristic index was obtained, and these indexes coincided with each other not only in physical meaning but also in quantity. Important conclusion drawn from these indexes was that the cavitation erosion mechanism is same in the incubation period as well as in the weight loss period, and that it is also same in laboratory testing apparatuses as well as in actual machines in the field. In the meanwhile, as to the erosion by solid particle impact, the concept of critical impact velocity was introduced to predict the behavior of solid particle at the impact on target material, rolling and skidding. It was made clear that the former causes the damage on the material by plastic deformation and the latter by cutting. This concept was useful to give rationale to the unexpected agreement of the material performance in the field with the result of laboratory test which was conducted under totally different experimental conditions from those in the field: in both cases damage was caused by the plastic deformation process only but without the cutting process at all. The following conclusion was obtained by integrating above research results on cavitation and solid particle impact that the common mechanism generating pure erosion damage in metallic material is plastic deformation.

Keywords: Cavitation attack, damage depth, incubation period, weight loss period, critical impact velocity, rolling, skidding, plastic deformation, cutting, characteristic index.

1. CAVITATION EROSION PROCESS

1.1. Time Dependency of Cavitation Damage Rate

The incubation period exists in the cumulative weight loss *vs.* time relationship for cavitation erosion process (Fig. (**1-8**)). In addition to this, the weight loss rate depends considerably on the testing time (Fig. (**1-9**)). These behaviors appear not only in the test result of the stationary specimen vibratory facility but also in those of the vibrating specimen facility and the water tunnel (Fig. (**1-4**)). This is quite inconvenient in the case, where cavitation-proofness of various materials has to be evaluated based on these test results. In the ASTM standard of testing method, it is required to continue the measurement until the S-shape curve as depicted in Fig. (**2-1(a)**) is obtained, and the order of merit in cavitation erosion resistance among materials is to be evaluated by comparing the so-obtained W *vs.* t relationship. According to this regulation of testing method, the mutual ranking of various materials may be obtained, but nothing more than that at all.

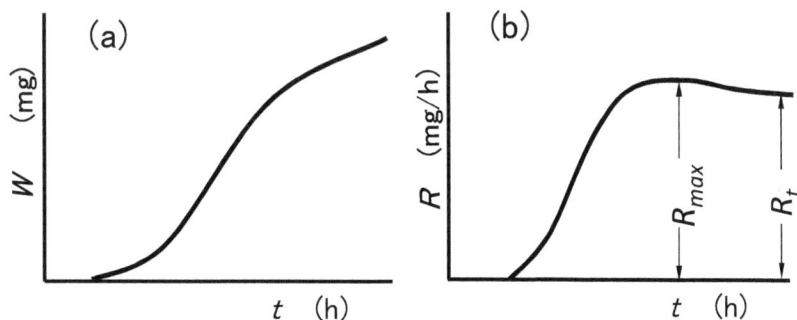

Fig. (2-1). (a) Cumulative weight loss of specimen, W, versus duration of time, t (b) Weight loss rate, R, plotted against time, t.

The peak level or the terminal level of weight loss rate, that is, R_{max} or R_t in Fig. (**2-1(b)**), may be useful for the evaluation of materials by single numerical value. However, question arises about the physical significance of these weight loss rates: the variation of R with t under the constant cavitation intensity might mean the transition of damage mechanism with time. The implication of this may be serious, when we consider the following case: the service life of some precision equipment may be terminated due to the increase in the surface roughness or the plastic deformation of the surface which might occur during the incubation period without any material separation or weight loss. In such a case, test data obtained in the process of severe material separation from the surface as shown in Fig. (**2-2**) might be of no significance.

Several parameters, including hardness and strain energy, have been proposed so far to represent the cavitation erosion performance of the materials, but none of single parameter is accepted as significant one. This may be the deserved consequence of the fact that the quantitative determination of erosion damage by single numerical value is not yet succeeded [1].

Fig. (2-2). Cavitation damage introduced on ARMCO iron specimen with vibrating specimen facility: frequency of vibration, 20 kHz; amplitude, 28 μm; test liquid, de-ionized water (40°C).

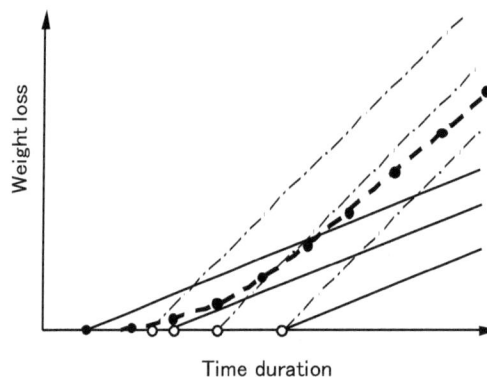

Fig. (2-3). Schematic reasoning for why weight loss *vs.* time duration relationship is a curve rather than a straight line [2].

The rational explanation for why the W vs. t relationship for cavitation erosion deviates from linearity was first proposed by Louis as follows [2]. A strict linear relationship ought to, theoretically, be established between weight loss and time. In spite of this, both material quality and cavitation intensity actually fluctuate over the material surface, and each element in the material surface accordingly separates independently after its own incubation period with its own weight loss rate, so that the weight loss *vs.* time relation for the whole test specimen, which is the sum of that of each element, must naturally become a curve (Fig. (**2-3**)).

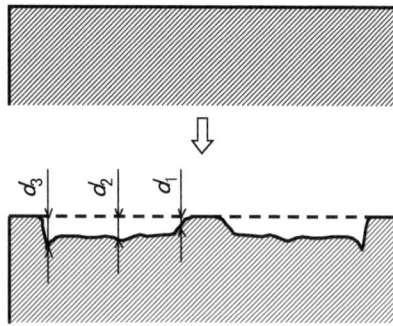

Fig. (2-4). Definition of damage depth, d [μm].

To substantiate his theory, it is required that the damage of each element in the material surface to be independently determined. To do this, Oka who was one of the collaborators of the author, adopted "damage depth" d [μm] as an index for the damage of element [3]: the corrugation which arose by the material separation in the test specimen surface was measured by using surface roughness meter, and then, it was determined in the way as shown in Fig. (**2-4**). It is, in physical significance, the volume of separated material per unit area of the surface with the dimension of [$\mu m^3 \mu m^{-2}$], which can be easily converted into the weight loss of the element.

The damage depth, d, was measured at five locations on a test specimen of water tunnel. They fell satisfactory on straight lines as shown in Fig. (**2-5**). It is also to be noticed in this diagram that the straight lines converged to a common point on the y-axis when they were extrapolated back to the beginning of test, that is, $t = 0$.

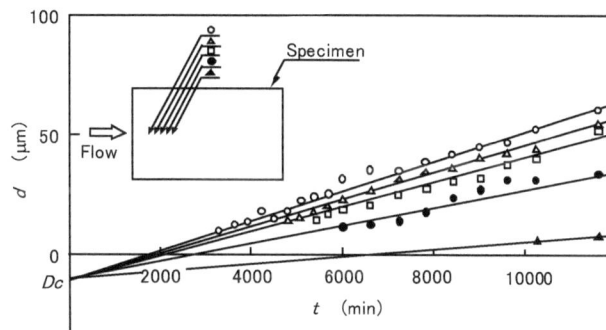

Fig. (2-5). Damage depth, d, versus testing time for a specimen of Type SUS 304 stainless steel tested in water tunnel [3].

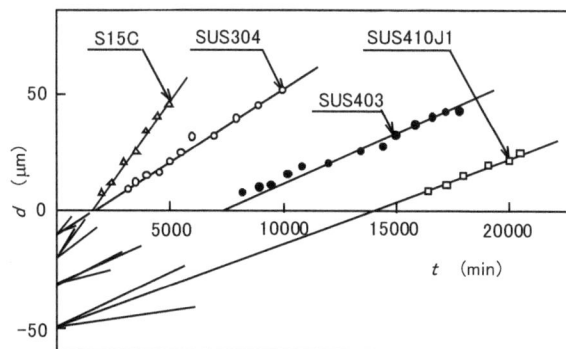

Fig. (2-6). Damage depth, d, *vs.* testing time for various materials tested in water tunnel [3].

The position of the point in the dimension of [μm], which was marked with D_C and named as "characteristic depth", was independent of the location on the specimen surface, where the damage depth was measured, but it was dependent on the material tested as shown in Fig. (**2-6**). Further, the position of D_C was not moved, even if the test equipment was changed from the water tunnel to the vibrating specimen facility and the stationary specimen vibratory facility, so long as the same material was tested (Fig. (**2-7**)). You may observe that the *d*-t relation obtained in water tunnel is broken, which should be naturally attributed to the shift of fracture mechanism of cavitation erosion at that time point. On the reason for the shift of mechanism, much will be discussed in Section 1.3 of the next chapter.

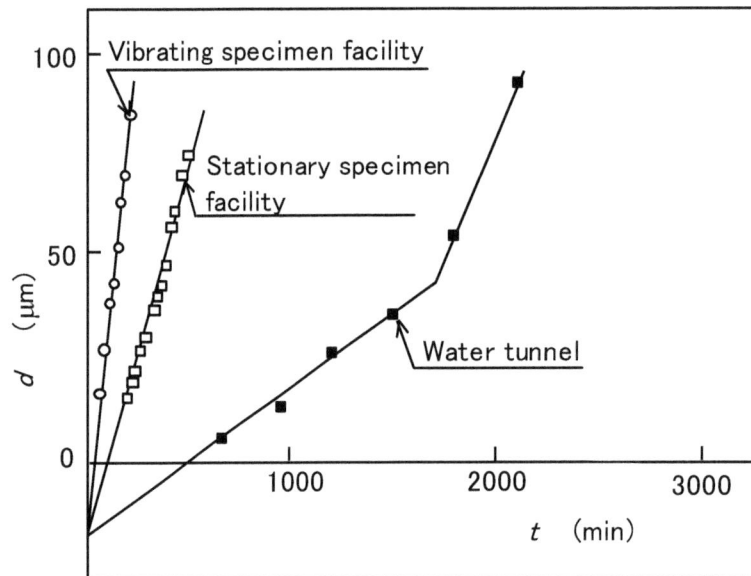

Fig. (2-7). Damage depth *vs.* testing time relationship for ARMCO iron in different testing apparatuses [4].

Now, in order to verify the theory of Louis which was given above, we compare Fig. (**2-5**) with Fig.(**2-3**). In the experiment, a linear relation was established between *d* and *t*, and that the incubation period, that is, the intersection of *d vs. t* line and *x*-axis, was dependent on the location of the specimen surface, in accordance with Louis theory. However, the gradient of the *d-t* line, that is, the weight loss rate of each element, changed not arbitral but in a sort of harmony with the incubation period, or, the longer the incubation period, the lower the weight loss rate was. This is nothing but the demonstration of the characteristic depth D_C. So, the conclusion that can be drown out from above comparison is that Louis theory is quit right in the prediction of cavitation intensity fluctuation but not so about material quality fluctuation.

The characteristic depth D_C importantly indicates that a certain significant process must be in progress during the incubation period, and that the rate of progress of this process is identical to the rate of material separation in the weight loss period. Besides, it also contains the notable possibility in the following points. Firstly, it proved that the damage mechanism in the accelerated testing facilities are same with that of the water tunnel which is generally considered to simulate well the cavitation erosion process of the real machines in the field (Fig. (**2-7**)). Secondly, it revealed that it is impossible for a single physical property of material to represent its erosion-proofness, because the damage of material depends on both the gradient of the *d-t* line and its intersection with *y*-axis, that is, the characteristic depth. On the other hand, it revealed, at the same time, a simple and convenient method to predict the damage depth on real machines: the characteristic depth, D_C, for the structural material can be easily determined by the short duration test in any accelerated testing facility. Together with this, one data of damage depth observed in real machine will give us the *d-t* line in the way as shown in Fig. (**2-8**), which can be utilized to predict the damage depth at any time point thereafter.

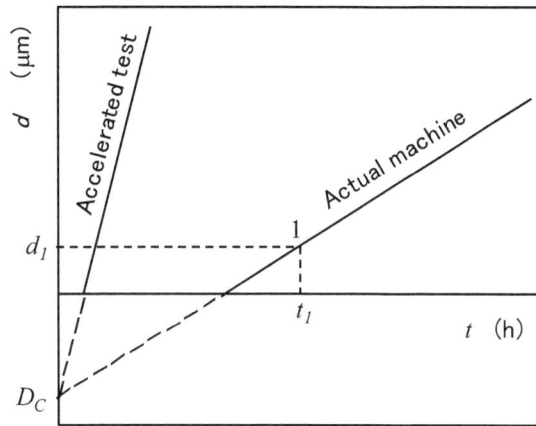

Fig. (2-8). A methodology for predicting depth of damage in service machines by use of D_C.

1.2. Physical Significance of D_C

The above mentioned methodology proposed for predicting damage depth d on the basis of D_C is concise and of practical convenience, but it is not completely without problems or limitations for application. First of all, we should remember that the reliability of this prediction procedure is entirely dependent on the validity of the physical significance of D_C. The most critical point concerning D_C is that it is a genuine material factor determined independently of cavitation intensity. Further, D_C represents d value at $t = 0$, and therefore, D_C must be correlated somehow to the material damage parameter, which has the dimension of length [μm]. Sakamoto who was one of the students of the author took notice of the hardness of the damaged surface, H_{VS}, which was estimated by extrapolating the H_V distribution profile on the cross section of cavitation damaged material (Fig. (**2-9**)).

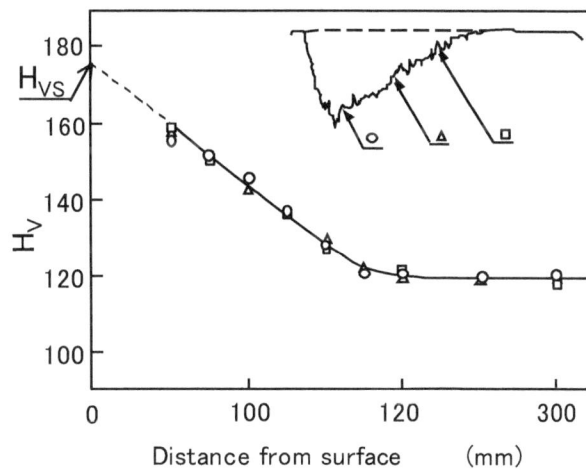

Fig. (2-9). Rise in Vickers hardness under cavitation damaged surface originated in water tunnel [4].

The level of H_{VS} so-obtained was naturally higher, as a result of the strain hardening, than at the deep inside of the material, where the influence of impulsive cavitation pressure did not reach. He was rather interested in the level of H_{VS} that was independent of the location of the specimen at which it was observed in spite of that cavitation intensity was expected to be different from location to location as described above. This might be the clue to reach the genuine material factor. So, he obtained the P *vs.* ε relationship for the material (ARMCO iron) of specimen by indenting it with a steel sphere of 3.17 mm in diameter (Fig. (**2-10**)), where P refers to the stress defined as load divided by projection surface area, $(\pi/4)d_S^2$, and ε is defined as d_S/D, where d_S is the diameter of the indentation and D that of the steel sphere.

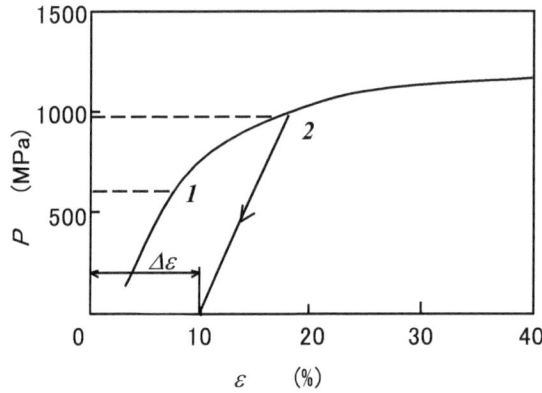

Fig. (2-10). Stress *vs.* strain (*P vs. ε*) diagram which was obtained by pressing a small steel ball on the surface of ARMCO iron specimen [4].

In fact, the stress *P* defined as such corresponds to Meyer hardness, and Mayer hardness and Vickers hardness have a unique correlation which is valid for well strain hardened material. Further, it is acknowledged that the initial hardness which is determined on the virgin surface (without strain hardening) corresponds to the yield strength P_γ of the *P vs. ε* relationship. Noting these features, he proposed an intelligent procedure for evaluating characteristic strain *Δε* as follows.

1. From P_γ (point 1 in Fig. (**2-10**)) and the measured H_V of the specimen at inside (Fig. (**2-9**)), proportionality constant *K* between *P* and H_V is determined.

2. Using the *K* value, P_C (point 2 in Fig. (**2-10**)) corresponding to Vickers hardness of the cavitation damaged surface, H_{VS}, is estimated.

3. From the point P_C in Fig. (**2-10**), a line parallel to the elastic part of the *P vs. ε* line is drawn and then the crossing point of this line with the horizontal axis gives the strain *Δε* concerned.

Fig. (**2-11**) summarizes D_C *vs. Δε* relationships obtained for five different materials. It can be seen here that the correlation between D_C and *Δε* is definitely significant. Thus, we may reach a conclusion that plastic deformation steadily accumulates during the incubation period, and when the plastic deformation has reached a sort of saturation point, the next stage of erosion begins, that is, the period of material separation or weight loss.

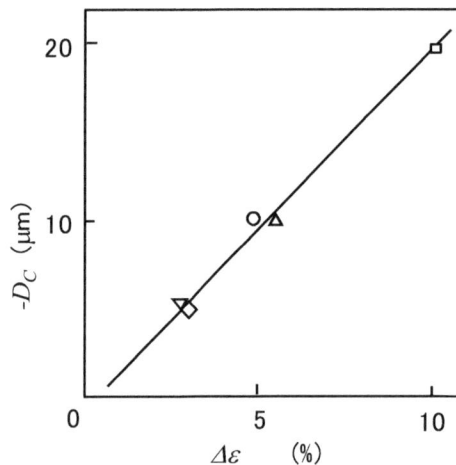

Fig. (2-11). Close correlation between D_C and *Δε* [5].

1.3. Erosion Process in the Incubation Period

The damage depth prediction procedure depicted in Fig. (2-8) is applicable only for predicting the behavior during the weight loss period, since d, damage depth, cannot be specified for the material during the incubation period. Thus, toward extension of the applicability of the cavitation erosion prediction method, it was desired to define some new index representing the extent of damage during incubation period.

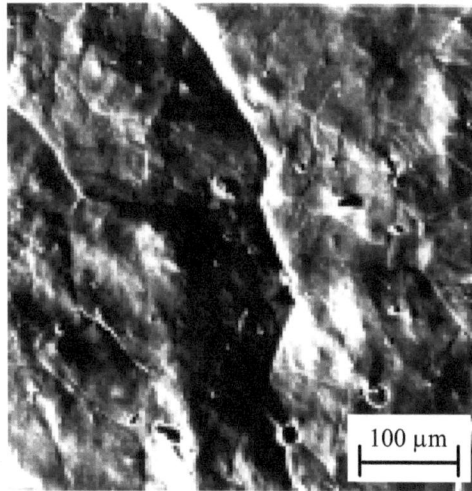

Fig. (2-12). Waviness and pits emerged on ARMCO iron specimen surface during incubation period prior to occurrence of material separation.

As shown in Fig. (2-12), the emergence of waviness and pits was observed over the ARMCO iron surface during the incubation period of the cavitation erosion. Yabuki, who was one of the collaborators of the author, proposed "surface increment percentage", ΔS, to represent the surface modification during the incubation period which was assessed with the following procedure: the initial effective surface length between two points separated by 1 mm along the horizontal direction of the specimen surface, that is, L_0 in Fig. (2-13) is measured using the surface roughness meter.

After a certain degree of progress of damage during the period, the effective length L between the same reference points is measured to calculate ΔS with the equation given in the figure. Fig. (2-14) summarizes the ΔS *vs. t* relationships in both logarithm graphs which were obtained for four different types of materials using the stationary specimen vibratory facility [6]. The following features are noted in this figure.

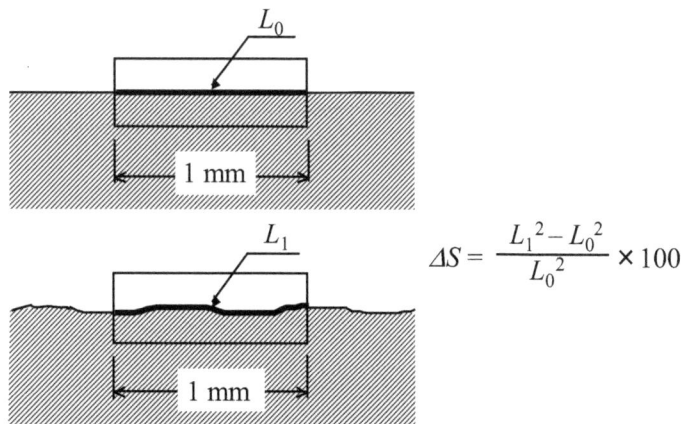

$$\Delta S = \frac{L_1^2 - L_0^2}{L_0^2} \times 100$$

Fig. (2-13). Definition of surface increment percentage, ΔS.

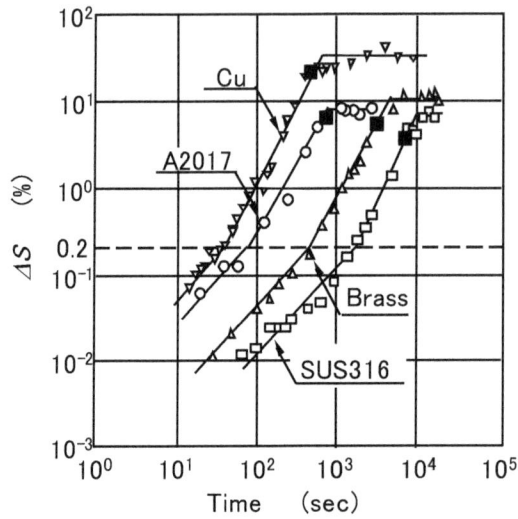

Fig. (2-14). *ΔS vs. t* relationships observed for different materials in stationary specimen vibratory facility [6].

1. Up to *ΔS* = 0.2, the slope of log *ΔS vs.* log *t* plot is 1.

2. After *ΔS* exceeds 0.2, the log *ΔS vs.* log *t* relation is still linear but the slope is greater than 1.

3. After reaching the end of the incubation period (indicated with ■ in Fig. (**2-14**)), the linear rise of log *ΔS* with log *t* continues for a while before reaching the constant level of *ΔS*. Similar measurements were conducted for brass specimen using three different types of testing facilities and the results obtained were summarized in Fig. (**2-15**).

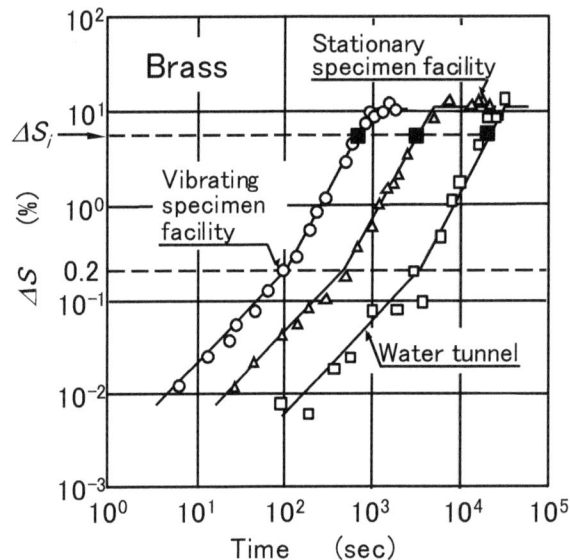

Fig. (2-15). *ΔS vs. t* relationships observed for brass specimen under different levels of cavitation intensity [6].

The *ΔS vs. t* relationships in both logarithm graphs were parallel to one another shifting along *t* axis depending on the cavitation intensity. It is worthwhile noting in the figure that the level of *ΔS* at the termination of the incubation period, which was designated as *ΔS$_i$*, was constant irrespective of the cavitation facility employed for the test, or the intensity of cavitation.

As depicted in Fig. (2-16), a procedure to predict incubation period by making use of ΔS vs. t relationships in both logarithm graphs was proposed: the first step is to obtain the log ΔS vs. log t relationship and determine ΔS_i by laboratory acceleration test for the material concerned; the second step is to measure ΔS_I of field service machine at the operation time t_I determining the point 1 in the graph; the third step is to draw straight lines crossing point 1 and parallel to that lines which were preliminarily determined by the laboratory acceleration test; then, the incubation period for the field service machine of the same material may be estimated as the intersection of so-drawn straight line and the horizontal line of ΔS_i.

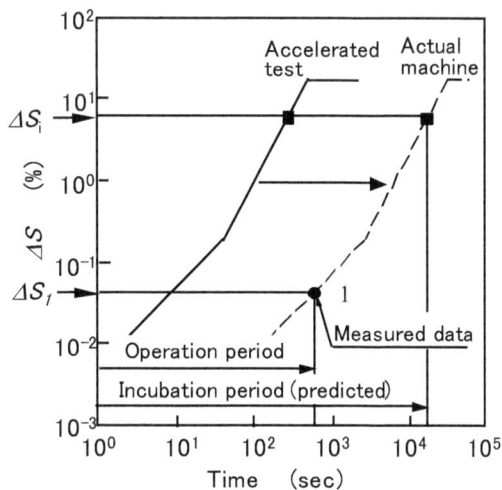

Fig. (2-16). Methodology for predicting the duration of incubation period [6].

When D_C was also preliminarily determined the slope of the d vs. t line in the weight loss period may be consequently obtained, and thus a single inspection would allow us to predict the extent of damage at any operation duration and further with a threshold extent the remaining life duration of the machine.

1.4. Physical Significance of ΔS_i

The above procedure for predicting cavitation erosion damage depth as well as remaining life duration is totally dependent on the validity of the estimated ΔS vs. t relationship. Thus, we felt it desirable to explain why the above observed ΔS vs. t behaviors would yield. For this, we looked into details of the relationship by dividing this relation into two stages, the first stage with ΔS smaller than 0.2 and the latter stage with ΔS greater than 0.2.

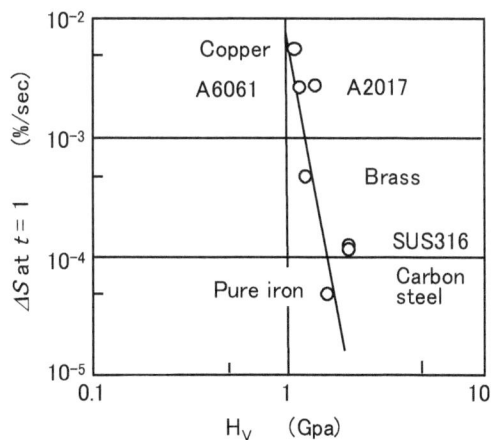

Fig. (2-17). Close correlation between slope of ΔS-t line in first stage and hardness (H_V) of the material [6].

As to the first stage, the correlation between the slope of ΔS-t line in a normal coordinate, that is the level of ΔS at $t = 1$ in Fig. (**2-14**), and the hardness of specimen material was examined (Fig. (**2-17**)), and a cavitation erosion model with the following features was adopted (Fig. (**2-18**)).

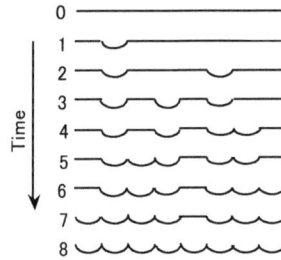

Fig. (2-18). Schematic illustration for the behavior of ΔS in the first stage [6].

1. Craters produced by cavitation impulsive pressure would not overlap before the entire surface is covered with them.

2. ΔS is 0.2 for a crater.

According to this simplified model, the behaviors of ΔS in Figs. (**2-14**), (**2-15**) can be easily rationalized as follows.

The reason why ΔS rises linearly with the time: The ΔS rise must be proportional to the coverage of the specimen surface with craters, which is proportional to the projected surface area of a crater as well as to the number of crater generated. Generation frequency of cavitation impulsive pressure which is high enough to produce crater would be constant independent of the lapse of time, and accordingly ΔS rise would be proportional with t. By the way, the lower the material hardness, the greater the area of the crater is, so that the rate of ΔS rise is in inverse proportion for the material hardness as shown in Fig. (**2-17**).

The reason why ΔS rise deviates from the linearity at the point of 0.2 to turn to more rapid rise: Provided that ΔS per single crater be 0.2 independent of the specimen material, ΔS at the time point of the entire coverage of the specimen surface with craters ought to be 0.2. This level of 0.2% is about 1/10 of the increment of the surface length which is originated when the indenter of Brinell hardness as well as Vickers hardness tester invades in the material. This may be due to the facts that the crater is actually three-dimensional or spherical and the needle of the surface roughness meter passes not always the deepest place.

In the examination of the latter stage of ΔS *vs.* t relationship, the correlation between the exponent of t (the slope of log ΔS *vs.* log t line) and the strain hardening exponent of material were examined to find a satisfactory correlation between them (Fig. (**2-19**)).

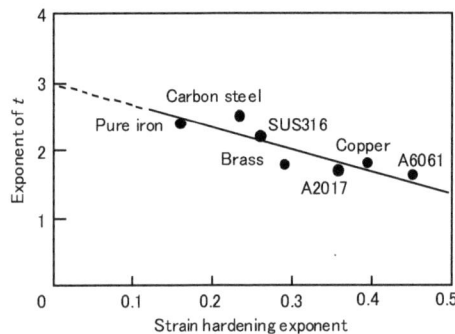

Fig. (2-19). Close correlation between exponent of t in the latter stage and strain hardening exponent for different materials [6].

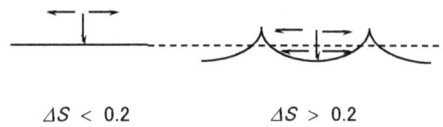

$\Delta S < 0.2$ $\Delta S > 0.2$

Fig. (2-20). Drawing which illustrates ΔS behavior in latter stage [6].

A model was adopted which clams that cavitation impulsive pressure would contribute more effectively in extending the curved surface of the latter stage than it extends the flat surface of the first stage (Fig. **2-20**). According to this simplified model, the behaviors of ΔS in the latter stage can be easily rationalized as follows.

The reason why the exponent of t is larger than that in the first stage: Next two factors, one accelerating, another suppressing, are concerned. (i) It would be more effective to extend the concave surface with edge than to extend the plane surface; with the lapse of the time, the crater is deepened and at the same time the edge rises from original plane; together with the effect of extending the plane surface, these two effects, one pushing the lower part of curved surface than the original horizon, another pushing the edge extruded higher than the horizon, must raise the exponent by three times. (ii) The spreading of crater raises the hardness of the material by the strain hardening effect; the crater consequently becomes resistant to be extended. As a result, the exponent index being lowered in proportion to the strain hardening index. Fig. (**2-19**) substantiate these factors mentioned above: the trend of the rising exponent with the diminishing strain hardening index; extrapolating the line of exponent of t vs. strain hardening index back to y-axis yields 3 for the exponent of t.

The reason why ΔS keeps rising still after the initiation of weight loss: The first separation of material must be at the summit of crater edge of the area which is restricted as compared with the whole area of crater; no significant change would be generated in the increasing rate of ΔS.

Lastly, the close correlation between D_C and ΔS_i seems to provide a definitive confirmation for the validity in the use of ΔS for the evaluation of cavitation erosion performance. As noted in Fig. (**2-21**), D_C and ΔS_i are not only in proportionality relation but also in comparable absolute values to each other for every given material.

In the earlier discussion, in the end of Section 1.2, on the physical significance of D_C, it was mentioned as "plastic deformation steadily accumulates during the incubation period", which is exactly the surface increment ΔS, and "a sort of saturation point" corresponds to "the termination of the incubation period". Namely, it may be concluded that D_C is exactly ΔS_i.

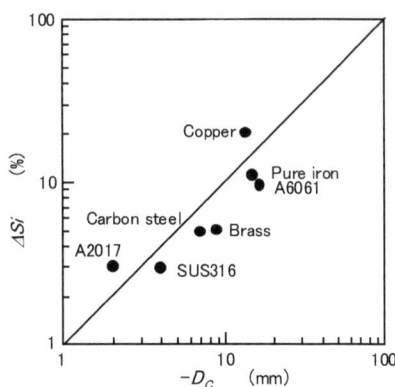

Fig. (2-21). Close correlation between D_C and ΔS_i [5].

In the end, it should be pointed out that the nature of cavitation damage during the incubation period is not similar to the metal fatigue phenomenon, where a crack propagates without plastic deformation of the outer surface, but rather to the constant-strain rate tension of a rod specimen, where the plastic strain is accumulated with the lapse of time until the rupture occurs. In other words, the elongation at rupture may correspond to ΔS_i of cavitation erosion process.

1.5. Application of Cavitation Erosion Prediction Method

Prediction of Cavitation Intensity in Service Machines: In the methodology proposed above for predicting cavitation erosion damage, the most tedious procedure may be the observation of ΔS or d in actual machine at least once during the service, which is inevitable for the determination of cavitation intensity in the machine. If it was determined just by monitoring the noise from the outer surface of machine, the prediction would be very simple and easy. This scheme is based on the following equation with the assumption that the parameters are independent to each other.

$$\text{(Rate of damage)} = \text{(Intensity of attack)} / \text{(Resistance of material)}$$

However, analysis of available cavitation erosion data indicated that such a simplified relationship is not valid in the real world. Fig. (**2-22**) summarizes damage rate R *vs.* "cavitation intensity" plots for five different metallic materials, where R was determined from the slop of the measured d *vs.* t relationship, and given in the unit of [μm min^{-1}] which can be easily converted to the engineering units of [mm y^{-1}].

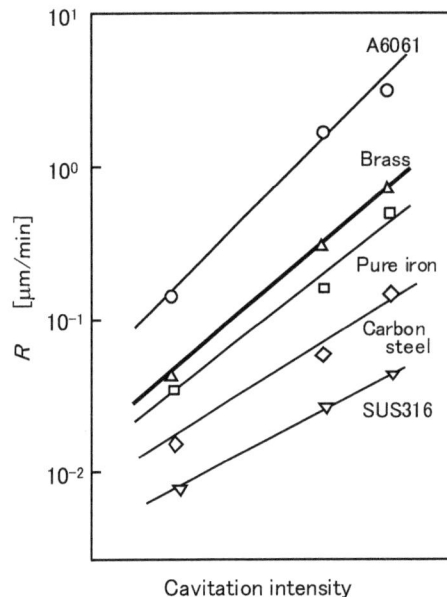

Fig. (2-22). Damage rate *vs.* cavitation intensity relationship for different materials [4].

In plotting the graph, the cavitation intensity in each testing facility was firstly sealed on the horizontal axis through the following procedure: two locations on the horizontal axis were arbitrary chosen; we took brass as the representative material and the higher damage rate of this material (obtained in the vibratory facility) and the lower rate (obtained in the water tunnel) were plotted at these locations. Then, a strait line was drawn crossing these data points, and the data point obtained in the stationary specimen facility was placed on this line to determine the location of cavitation intensity in the facility on the horizontal axis. Thus, the cavitation intensity in each testing facility was determined on the horizontal axis. Data points for the rest four specimens of various materials were placed at adequate cavitation intensity positions. Then, as seen in the figure, each of them fell on individual linear relation ensuring the validity of the relative intensity of the testing facility.

One of the remarkable aspects in the graph is that the lines are not parallel with each other but the interval between R levels at the higher cavitation intensity side is wider than that at the lower intensity side, implying that the resistance of material against cavitation attack is not constant but different depending on the cavitation intensity (remark the vertical axis being logarithmic scale), the reason for which is given as follows. As can be seen in Fig. (1-3), the cavitation intensity of testing facility must be given by the integration of different level of impulsive pressure with its generation frequency. So that, lower cavitation intensity implies lower generation frequency of higher impulsive pressure but those of lower impulsive pressure remain unchanged. As a consequence, the damage to the material of higher resistance will be remarkably reduced but that to the material of lower resistance will be changed not so much, which must appear that the resistance of the material raised at lower cavitation intensity.

This argument indicates that the equation given above is not valid, and therefore it is essential to measure the cavitation intensity and the resistance of the material simultaneously. In fact, such simultaneous measurement is realized in the open inspection of field service machines during shut-down period.

Properties Representing Erosion Performance of Material: The material of favorable erosion performance is the material endowed with low erosion rate, or the material with low rate of ΔS rise. In other words, the ΔS *vs.* t line in the stage of ΔS lower than 0.2 is preferably positioned more on the right-hand side in Figs (2-14) and (2-15) and the slope of the line in the stage of ΔS greater than 0.2 is as small as possible. The position of ΔS_i determines the duration of the incubation period but has nothing to do with the damage rate. For the ΔS *vs.* t line to be located more on the right-hand side, the material should be endowed higher hardness (Fig. (2-17)). On the other hand, for the smaller slope of the ΔS *vs.* t line, the strain hardening index ought to be greater. Thus, the material property which represents the erosion-proofness is hardness and strain hardening index. In fact, these requirements derived from the laboratory tests are exactly the same that stated in 1987 by Ozaki and Onuma who developed in the field a new type of stainless steel with excellent resistance to cavitation erosion [7].

2. EROSION BY SOLID PARTICLE IMPACT

2.1. Research History

Solid particle impact erosion is sometimes called erosive wear, and classified into the wear phenomenon together with abrasive wear, adhesive wear and fatigue wear. It has, therefore, widely been recognized that the nature of this sort of erosion is pure mechanical process, and the assertion is possible that there is completely no relation with corrosion. In this point, the research history may be different from the case of cavitation erosion.

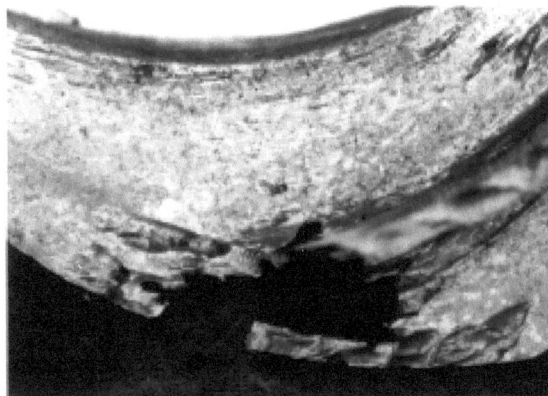

Fig. (2-23). Appearance of pump impeller attacked by corrosive slurry containing gypsum particles in stack gas de-sulfurization plant.

Of course, erosion by solid particle impact and corrosion is sometimes generated simultaneously in the metal surface. Rather, frequently reported are the cases of serious damage to instruments in the field which occurred in corrosive slurries, that is, the mixture of corrosive solution and solid particles. So in this section, this sort of combined damage of erosion and corrosion is named slurry erosion corrosion and discussed in detail thereafter.

In 1970's the author encountered the slurry erosion corrosion problem, implicated with conservation of environment. Serious damage occurred on the pump components, which circulated cleaning liquid to remove sulfur compounds from the stack gas of thermal power plant. This cleaning liquid was of high acidity containing solid particles of gypsum. It seemed to the author nothing but the synergistic effect of erosion and corrosion at that time, when the pump impeller made of the tough alloy with hardness over H_V 200 was cut by the impact of gypsum particles of H_V 20-50. In fact, the impeller had to be exchanged after the operation of several years (Fig. (**2-23**)).

In order to cope with the slurry erosion corrosion, as described above, appropriate material selection and quantitative prediction of the damage which the material suffers are indispensable. To achieve these aims, slurry erosion corrosion tests must be carried out on various materials. Serious problems, however, exist in the tests: firstly, it is not easy to conduct slurry erosion corrosion test in laboratory as it is described below, and secondly such laboratory test results seemingly do not necessarily agree with the performance of the material in real machines in the field, which will be discussed in the next section.

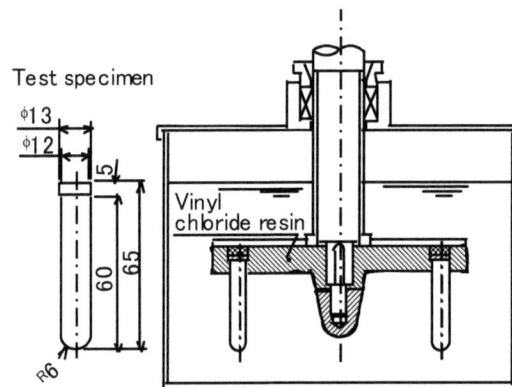

Fig. (2-24). Slurry pot type testing apparatus [9].

As an example of the first problem mentioned above, referred below is the result of slurry erosion corrosion test which the author and collaborators carried out using a device similar to Stauffer's Grinding Pot [8], in that the finger-shaped test specimens were driven around in a round slurry tank (Fig. (**2-24**)). With this device, the weight loss scarcely occurred in the test of short period. Even after a long testing duration the test results scattered badly, since the amount of weight loss was too small as compared with the weight of specimen. These troubles were attributed to the fact that the slurry in the tank ran around with almost equal velocity to that of the specimen. Installation of baffle plates in the tank to suppress the rotary motion of slurry, however, caused unexpected results: an increase in the weight of the specimen occurred, or, the weight loss of the specimen under erosion corrosion condition was smaller than that under pure erosion condition. The root cause of this unacceptable test result was that erosion corrosion area and corrosion area occurred separated on the specimen surface: the front side surface of the specimen was eroded as well as corroded by slurry stream, while the rear side, where solid particles of slurry scarcely impinged, was solely corroded resulting in the formation of a thick corrosion products layer and even some deposits which must have suppressed the weight loss of the specimen. No significant interpretation could be put on the weight loss of the test specimen. We had to conclude that the device is not suitable for an erosion corrosion testing method [9].

In order to avoid such trouble, how about conducting slurry erosion corrosion test by flowing slurry through the water tunnel which is usually used for cavitation erosion tests? In the test specimen surface, it is naturally expected that the erosion area would coincide with the corrosion area. The answer would be regretfully negative, because of the pump impeller: the tip of impeller is usually rotating with higher velocity than that of slurry on the test specimen so that it must be damaged sooner than the specimen would be.

Then, how about holding slurry in a closed tank, and extruding it not with pump but with compressed gas through pipe line installed with a nozzle at the end to impinge the slurry jet with high velocity on specimen surface? In fact, such equipments as shown in Fig. (**2-25**) were manufactured actually in the 1980's when advanced fossil fuel technologies, such as coal liquefaction, tar sand processing, required the handling of slurries in the conditions conductive to the combined erosion and corrosion. For example, the direct coal liquefaction process involved pressure letdown valve with an orifice where the stream velocity of corrosive slurry containing typically 10% ash could be in the several decade meter per second.

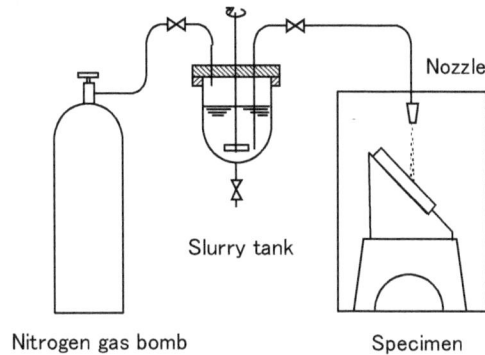

Fig. (2-25). Schematic drawing of slurry erosion test apparatus of high pressure gas driven type.

In several runs at the beginning of operation, reliable data were obtained smoothly. But, soon it became difficult to maintain the velocity of slurry jet at constant since the nozzle bore was widened by the flow of slurry in spite of that it was made of hard ceramics. And then, serious wall thinning occurred at the bend of the pipe line, and the corrosive slurry spouted out through the opened hole fouling all over around. The equipment was scrapped soon.

2.2. Parameters Affecting Solid Particle Impact Erosion

It has often been experienced by designers and engineers that any experimental result obtained by those laboratory erosion tests could not be applied to the real machine. This was due to the discrepancy between the result of laboratory test and the performance of the material in the field, which was to be attributed firstly to the fact that the extent and the nature of the damage in the metal surface which was caused by the solid particle impingement strongly depended on relating parameters, and secondly that there were too many strongly influencing parameters as follows: the physical property of target material (hardness, toughness), the property of solid particle (size, shape) and the impact condition of the particle (impact speed, angel).

Fig. (**2-26**) shows typical dependencies of erosion rate on impact angle for ductile material (mild steel, aluminum *etc.*) and a brittle material (quenched steel, ceramics *etc.*). Here erosion is defined as the mass of removed target material per unit mass of impinging particle.

It should be noted that the ductile material exhibits the most severe damage at impact angles lying typically in the range 20° to 30°, which is called cutting process, and rather mild damage at high impact angles, which is called deformation process. In contrast with this, the brittle material is damaged by cracking process only with the simple impact angle dependency: most rapidly under conditions of large impingement angle, and minimally at shallow angles. Suppose particles were directed at a right angle against the specimen in laboratory test, brittle material would suffer heavily and ductile material slightly. This test result will nominate the ductile one for the construction material of real machine. However, if it was used for a pneumatic powder transport pipe, where solid particles impinge at shallow angel, lager damage will result on the ductile material than on the brittle one. Thus, from this example it can be seen that the performance in the field is often the opposite to the indication of the laboratory test.

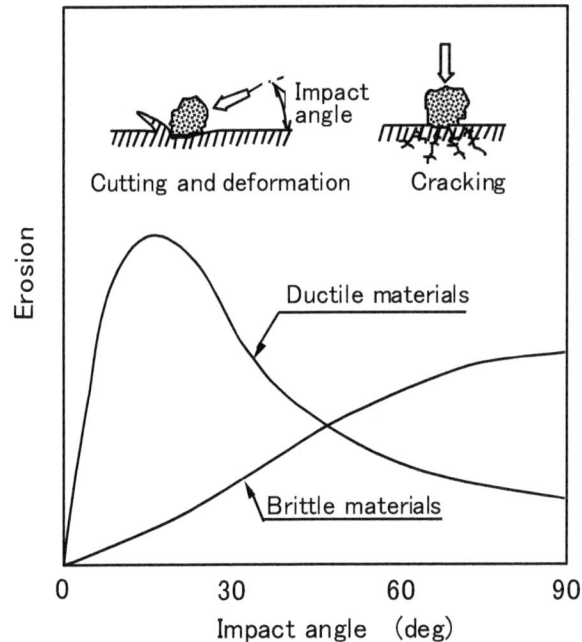

Fig. (2-26). Schematic representation for erosion rate *vs.* particle impact angle relationship for brittle material and ductile material.

Then, how about making the particle impact angle in laboratory tests coincide with that in real machines: this would be successful provided that the impact angle in the real machine could be precisely predicted. The problem with this is that even in laboratory machines the particle impact angle is not clear, since it is not necessarily coincident with that of the fluid flow in which the particles are suspended. It will be surely much harder to predict it in real machines.

Even if the prediction or the measurement of impact angle were successful in both the machines, the test result will still be far from predicting the real damage so long as exactly the same particles with those in the field were not used in the laboratory test. Fig. **(2-27)** shows the results of two series of erosion tests which were conducted by Sheldon [10], under the same impact velocity on the same group of materials but using silicon carbide (SiC) particle of two different sizes: grit mesh 120 (particle size of 127 μm in mean diameter) *vs.* 1 000 (9 μm).

If we make a comparison between the hardened steel and the plate glass in their resistance to erosion caused by the impact of 127 μm particles (Fig. **(2-27(a))**), the former will be naturally judged more resistant to erosion than the latter because it sustains a smaller erosion at every impact angle. In contrast with this, in the erosion test using 9 μm particles the complete variation of erosion manner of the same materials is recognized: all the materials behaving in a ductile or more ductile manner in which erosion reached a maximum at an angle considerably less than 90 deg, and the hardened steel sustaining a far larger erosion over all impact angles as compared with the plate glass (Fig. **(2-27(b))**). Thus, the conclusion from the result of laboratory test using the particle of 127 μm would be the very reverse of the performance of the materials in real machine if it was engaged with 9 μm particle. Some larger particles than those used in the field are often adopted for laboratory tests in order to shorten the test duration through accelerating the erosion rate of test specimen. The test results obtained under such accelerated test conditions cannot be extended to real machines even if the measurement were conducted at every impact angle.

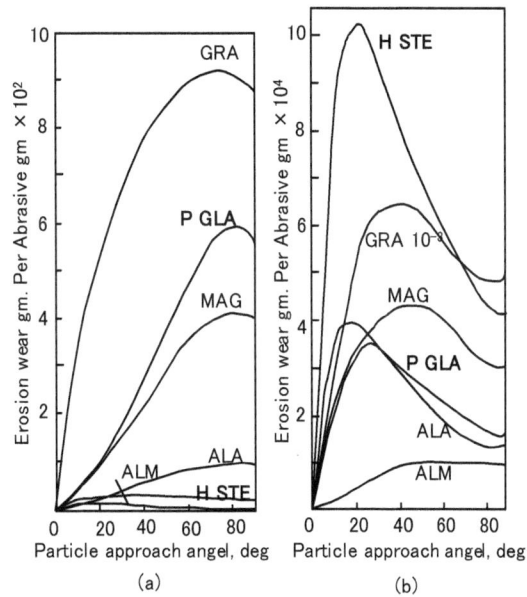

Fig. (2-27). Effect of particles size on erosion: (a) Predominantly brittle manner; grit mesh size, 120; (b) Predominantly ductile manner; grit size, 1000: Parameters in common: particle impact velocity, 150 m sec^{-1}; particle, grit of SiC; material; GLA, graphite; P GLA, plate glass; MAG, high density magnesia; ALM, annealed aluminum; ALA, high density alumina; H STE, hardened steel [10].

Then, how about using the particles obtained in the field for laboratory tests, and measuring the erosion rate at every impact angle? The answer is negative, because under such test conditions the erosion rate of test specimens will be reduced to nearly the same level as those in the field, and far longer test duration, say weeks or months possibly, will be required to cause some measurable weight losses to the specimen. This is not an exaggeration of the problem; the reduction of the particle diameter from 127 to 9 μm resulted in the damage rate being reduced to one hundredth (compare the scaled of ordinates in Fig. (2-27)). Particles impact velocity in laboratory tests ought to be unavoidably raised in order to obtain results within the limited duration of test. However, it has an influence of no less importance than particle size does, not only on the extent of damage but also in the brittle/ductile manner of target material. Fig. (2-28) is an experimental result obtained by Hojo and his colleagues about the influence of particle impact velocity on the erosion extent and the brittle/ductile manner of erosion damage of poly-methyl-methacrylate resin.

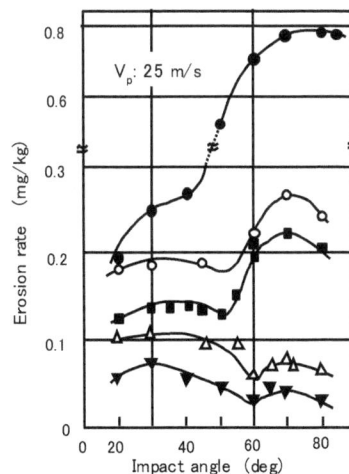

Fig. (2-28). Effect of impact angle on erosion rate of poly-methyl-methacrylate resin with glass beads (170 μm in mean diameter) impinged at various impact velocities [11].

At lower impact velocities, the maximum of erosion at shallow impact angles was larger than that at high angles, which indicated the ductile manner of the resin. On the other hand, at higher impact velocities, the maximum of erosion at high impact angles was larger than that at shallow impact angles. This result indicated a typical variation for a single material from ductile to brittle manner depending on particle impact velocity [11]. This variation of erosion manner may further depend on the characteristics of the material, namely will be different from one material to another. Therefore, the result of laboratory erosion test which was conducted with the particle impact velocity different from that in the actual machine cannot be expected to coincide with the performance of the material.

It may be concluded as follows. The erosion extent and brittle/ductile manner of erosion damage depend strongly on the impact angle, the size, the impact velocity of particle as well as on the characteristics of target material. This leads to the realization that any result of laboratory erosion test on materials, even the ranking order of merit, cannot coincide with that of their performance in the field, provided that all of these experimental conditions were not precisely coincide with those of the actual machine. A laboratory test, however, which is conducted under the same conditions with those in the field, is no more laboratory test but real machine test.

2.3. Characteristic Surfaces in a Crater

In the preceding section, it was demonstrated that solid particle impact angle exerts decisive effect on the erosion rate irrespective of the ductile/brittle manner of erosion. The author and the collaborators examined its effect on corrosion by measuring dissolution rates of characteristic surfaces in the crater of large dimension which was formed on commercial pure iron specimen by allowing the projectile, a cylinder with the rod head of 4 mm in diameter, to fall freely on it (Fig. **2-29**).

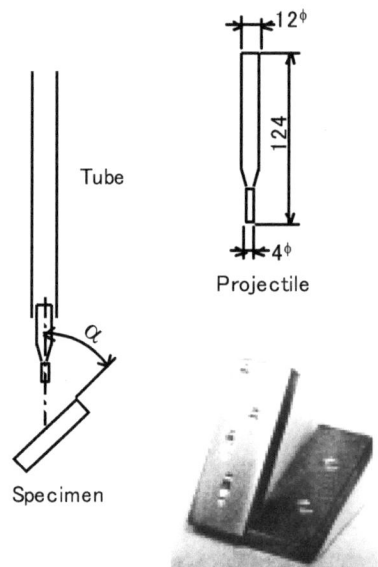

Fig. (2-29). Schematic sketch of free-falling projectile [12].

Careful observation was performed on the shape of crater cross-section to find characteristic surfaces as shown in Fig. (**2-30**): (1) displaced surface; (2) lip surface and (3) surface produced by sliding; in the crater formed with the impact angel of 50°, and (4) cut surface in the crater with the impact angle of 30°. Dissolution rates of the characteristic surfaces were determined by measuring the weight loss of a specimen dipped into a hydrochloride acid solution of pH 2: each specimen had a few craters formed with the same impact angle on its surface and was masked completely by an anti-corrosion paint except for one of the characteristic surfaces in each crater. The results are shown in Fig. (**2-30**) as the function of duration of exposure to the acid solution.

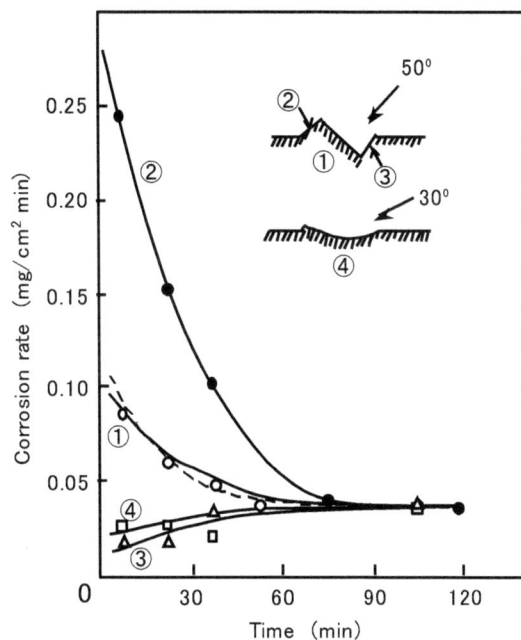

Fig. (2-30). Characteristic surfaces and the dissolution rate: (1) displaced surface, (2) lip surface, (3) sliding surface, (4) cut surface [12].

The dissolution rate of original polished surface without crater is also given with the broken line. All of the dissolution rates changed with time to converge upon a certain common level. This might result from that the dissolution had removed surface material until the depth was reached where no further influence of particle impact or of polishing was observed. The initial dissolution rate of characteristic surface was obtained by extrapolating the curve back to the initiation of corrosion test. The so-determined dissolution rates of the polished surface (broken line) and the displaced surface (1) were nearly the same at about 0.10 mg cm^{-2} min^{-1}. The rate for the lip surface (2) was approximately 0.3, which was tree times higher than that for the polished surface. In contrast, the rates for the sliding surface (3) and the cut surface (4) were both about 0.02, which was only one fifth of that for the polished surface [12].

The question might then arise as to why the lip surface (2) dissolves at a rate fifteen times higher than that for the sliding surface (3) or the cut surface (4). This might be attributed not only to the corrosion accelerating agents of the microscopic as well as macroscopic defects originated on the lip surface (2) which accompanied the formation of the characteristic surface, but also to another important cause which reduced the dissolution rates of surfaces (3) and (4) to one fifth of that for the polished surface.

Observation was carried on SEM micrographs of etch pits formed on the different characteristic surfaces. On the polished surface, each pit took different form depending on the crystal grain on which it was located. This suggested that the polished surface consisted of various crystal planes. A similar situation was observed on the displaced surface (1), as well as on the lip surface (2). In addition to this, a larger number of pits were formed on the lip surface (2), indicating the presence of many microscopic defects. Macroscopic defects such as swelling and cracking were also observed. In contrast to these, all of the pits on the cut surface (4) assumed regular square shape arranged in a common direction irrespective of the crystal grains. The same was observed on the sliding surface (3). Moreover, the inner facet of each pit assumed a common shape. This appeared to prove that the crystal plane {110} corresponded to those surfaces and that the projectile sliding direction was <111>. These are characteristic slip plane and slip directions of a bcc metal. Also, it had been well established that among crystal plane of bcc metal the dissolution rate of {110} plane is second lowest, the lowest being {100} plane. Accordingly, this particular crystal structure seemingly has caused the sliding surface (3) and cut surface (4) to dissolve at a lower rate than that for the polished surface.

It has been made clear that under the condition in which erosion and corrosion are being generated simultaneously, a slight change in the particle impact angle must cause a great change in the corrosion rate of the target material. So, we cannot but conclude that as compared with the case of pure erosion it must be much more difficult for laboratory test results to agree with the performance of materials in the field.

2.4. Laboratory Test and Performance in the Field

Jet-in-Slit[1] Testing Apparatus for Slurry Use: This testing apparatus was developed with the concept that it is more important to continue operation than simulating the erosion in the field. The main structure and the details of apparatus are shown in Figs. (**2-31**) and (**2-32**): the main tank of transparent polyvinyl chloride resin consisted of two sections placed one over the other. In the lower section of smaller diameter, a fluidized bed was set up. In the upper section of the tank, the slurry exhausted from the test section was separated into solid particles and clear liquid.

Solid particles precipitated to fall down into the fluidized bed below. The bulk of the clear liquid was circulated by pumps into the test section and rest into the underside of the distributor to set up the fluidized bed where solid particles were suspended uniformly in the liquid flowing upward. The liquid from nozzle (1.6 mm in diameter) located in the center of the test section sucked up the slurry, which was mixed with the jet liquid to impinge upon the surface of the test specimen (18 mm in diameter by 4 mm high) and thereafter exhausted radially through the slit between the specimen and guide plate. The damage occurred at the area of jet impingement on the specimen as well as on the surface outside this area, which was due to the radial flow. Four pieces of the test section were installed in the tank. The operation conditions of the apparatus were as follows: slurry impact velocity, 1.7 m sec^{-1}; impact angle, 90°; flow rate, 2 L min^{-1} [13].

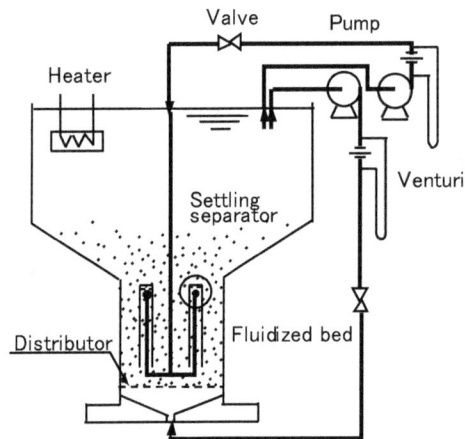

Fig. (2-31). Schematic diagram of jet-in-slit testing apparatus applied for slurry erosion corrosion test [13].

This testing equipment has, though the slurry impact velocity (1.7 m sec^{-1}) is remarkably lower than those in real machines, excellent advantages: (1) the damage develops only on the testing surface of the specimen; no other part of the apparatus is damaged; (2) the amount of particle, specimen size, and the power consumption are all small; and (3) the reproducibility of the test results is excellent. The most important feature is that the entire test surface is subjected simultaneously to erosion and corrosion. This is indispensable for obtaining reliable test results as described in the preceding section.

Preliminary tests on Type 316L stainless steel with corrosive slurries of silica sand resulted in the specimen weight loss of sufficient quantity to be measured at high accuracy in comparatively short duration of testing time. Furthermore, fairly linear relations existed between the volume loss of the specimen and testing time proving good reproducibility of test results (Fig. (**2-33**)).

[1] On jet-in-slit refer the detailed description in Chapter 4.

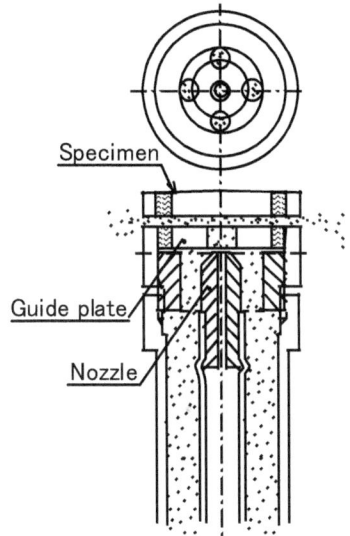

Fig. (2-32). Enlarged view of test section [13].

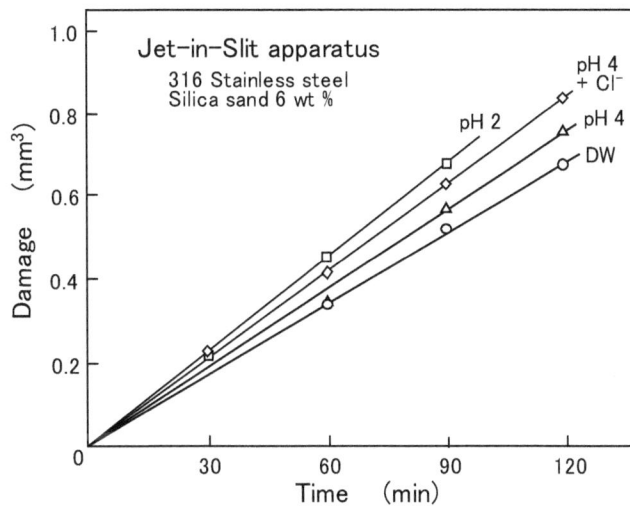

Fig. (2-33). Slurry erosion corrosion damage versus testing time for Type 316 stainless steel: slurry, silica sand (6 weight %) and respective solutions; slurry impinging velocity, 1.7 m sec^{-1}; slurry temperature, 60 °C [13].

Slurry erosion corrosion tests were conducted on Material A (H$_V$ 280) and Material B (H$_V$ 500) which were of high Ni and Cr content to be used for slurry pump components. Stainless steel of Type 316L (H$_V$ 198) and Hastelloy C (H$_V$ 210) were also included in the tests as reference material. Slurries were prepared from the corrosive liquids of nearly same chemical compositions as those in an actual stack gas scrubber, and the de-ionized water (DW) for the reference slurry. As to particles, the gypsum (H$_V$ 20-50) particles which were the same that contained in the slurry of the scrubber were used. Silica sand (H$_V$ 1200) particles were also used for the reference slurry.

The results of test are given in Fig. (**2-34**) for silica sand slurry and in Fig. (**2-35**) for gypsum slurry. It should be remarked that the damage rates originated with the gypsum slurry were extremely low as compared to those in the silica sand slurry. This might be attributed to the size as well as to the hardness of the particle. As to the order in damage rate, stainless steel and Hastelloy have exchanged their position, which we may take rather natural. But unexpectedly, Material A and B stayed in the same position. Furthermore, their relative damage rates agreed with each other. All the specimen surfaces showed similar matted appearance.

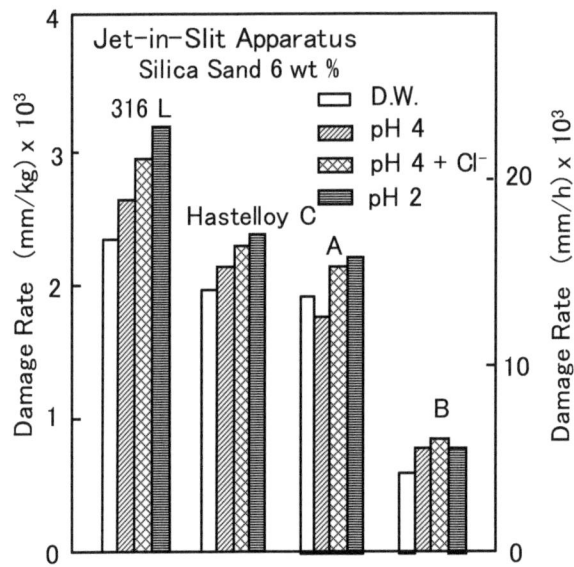

Fig. (2-34). Damage rates of materials in jet-in-slit apparatus: slurry, respective liquids and silica sand (6 weight %); slurry impinging velocity, 1.7 m sec⁻¹ [13].

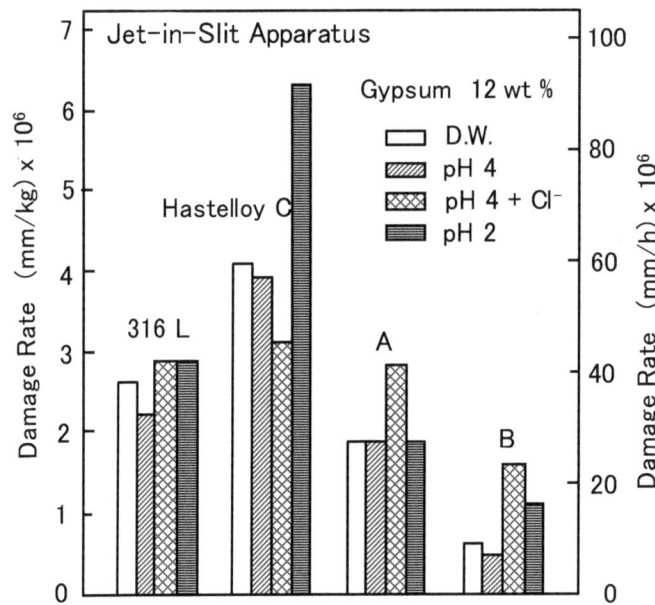

Fig. (2-35). Damage rates of materials in jet-in-slit apparatus: slurry, respective liquids and gypsum (12 weight %); slurry impinging velocity, 1.7 m sec⁻¹ [13].

Performance in Slurry Pump: A good opportunity was given to verify whether the test result obtained in laboratory can be applied to real machine in the field: two slurry pumps in the flue gas de-sulfurization plant had to be renewed [13]. Material A and B was used for the components of Pump A and B, respectively, which had common dimension and performance: bore diameter, 125 mm; impeller diameter, 265 mm; revolution, 1750 rpm; discharge rate, 85 m³ h⁻¹. They pumped gypsum slurry of sulfuric acidity at the same time under nearly the same conditions: the solid particle concentration ranged between 20 and 40 weight %, and pH value between 2 and 4. The average depth of the damage on the blade top of impeller and the surface of casing was measured on regular intervals. Fig. (**2-36**) shows the result.

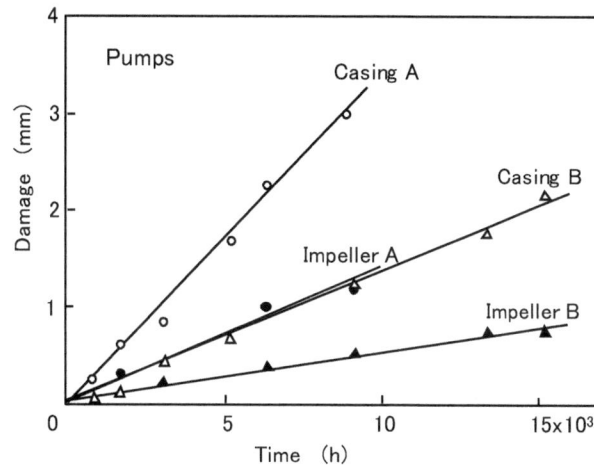

Fig. (2-36). Gypsum slurry erosion corrosion damage versus operation time for pump components in actual service [13].

It should be recognized that the damage depth increased linearly with the duration of operation just as in the case of laboratory test. In each pump, the casing was damaged more deeply than the impeller, and the damages of Pump B were one half or one third smaller than Pump A, which is in good agreement with the test result in Figs. (**2-34**) and (**2-35**). However, this is, surprisingly, a result of the very opposite to the content that argued in preceding sections.

2.5. Critical Impact Velocity in Solid Particle Impact

Concept of Critical Impact Velocity: The unexpected agreement of the performance of the materials with the test result described above occurred in spite of the great difference in the related conditions. Precisely, the particle impact velocity in laboratory test was 1.7 m sec^{-1} and in the slurry pump it must exceed at least 10 m sec^{-1}. The particles impact angle in laboratory test was apparently 90° and in the actual slurry pump it must be very shallow. In particular, in the laboratory test using silica sand slurry, not only the size but also the hardness of the solid particle was completely different from those in the gypsum slurry pumps: the mean diameter of particle, 77 μm for silica sand *vs.* 40 μm for gypsum; the hardness of particle, H$_V$ 1200 for silica sand *vs.* H$_V$ 20-50 for gypsum.

A possible clue to the rationale of this unexpected agreement may be the impact velocity: with reducing impact velocity, solid particle would tend not to skid on the target material surface, which must accordingly decrease the damage by cutting process; at a certain lower velocity, the particle would not skid any more, which would result in no cutting damage at all but the deformation damage only. Naming this velocity as "critical impact velocity", it might be interpreted as follows: in the laboratory test the impact velocity, that is, 1.7 m sec^{-1} must be too low and the impact angle, that is, 90° too high for the particle to skid at the impact, resulting no cutting process at all. In other words, the critical impact velocity for the particle impact conditions must be far higher than 1.7 m sec^{-1}.

On the other hand, the solid particle in the slurry pumps was smaller in size and lower in hardness than the particle used in the laboratory. Even if these particles impinged on the metal surface with a high velocity at a low impact angle, the material would not be damaged by cutting. In other words, the critical impact velocity in the slurry pump must be higher than the actual impact velocity. Then, it must be a matter of course that the results of the laboratory test agreed with the actual performance in slurry pumps because the damage mechanism was commonly the deformation process only in both cases.

To provide the proof of the critical velocity, an experiment was carried out by allowing silicon carbide particles to fall freely on the mild steel surface which was inclined at a certain angle [14]. The specimen with the damaged surface was put into a hydrochloric acid solution of pH 2 and its dissolution rate was measured. The dissolution rates of surfaces which were damaged at various impact angles gradually converged upon a certain common rate with reducing impact velocity as shown in Fig. (**2-37**).This impact velocity of 1.7 m sec^{-1} at which dissolution rate came to a common value must be the critical impact velocity because unity of dissolution rate means the unity of damage mechanism.

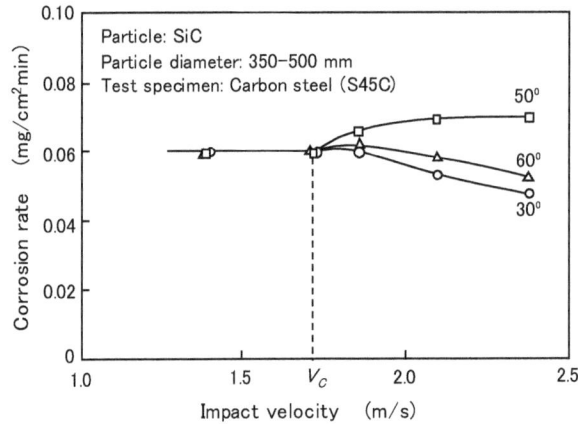

Fig. (2-37). Relationship between dissolution rate of damaged surface and solid particle impact velocity on mild steel [14].

Principle of Critical Impact Velocity Measurement: It would take a lot of time and work to determine the critical impact velocity for the various materials in various slurries by measuring the dissolution rates of the damaged surfaces. So, the methodology to determine the critical impact velocity through measuring the friction coefficient was established basing on the following principle: during the solid particle impact on the target material surface, the motion of the particle is affected by the friction force in the tangential direction; dynamic friction force when it skids, and static friction force when it does not skid but rolls. It is well known that dynamic friction coefficient is smaller than static friction, so that we can judge with the level of friction coefficient whether the solid particle skids or rolls on the surface and accordingly whether the cutting process is included or not.

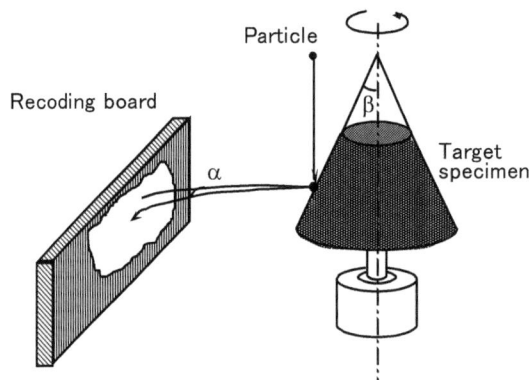

Fig. (2-38). Schematic sketch of rotating target apparatus for measuring friction coefficient during particle impact [14].

The rotating target apparatus in Fig. (**2-38**) was used to determine the friction coefficient during particle impact. A solid particle is allowed to fall from a certain distance down to impinge on the rotating conical target specimen. After the impact, the solid particle rebounds off toward the measuring board. During the impact, the solid particle is affected by a friction force in the tangential direction due to the rotation of the specimen to cause some deviation in its rebound direction. The deviation in angle is determined through the distance between two points which are marked by the particle impact with and without the specimen rotation. The friction coefficient is derived from the angle of deviation, α. The particle impact velocity is controlled through the falling distance, and the impact angle through the half vertical angle, β, as well as the specimen velocity.

During the impact of particle on target surface, in other words during the contact of the particle and the target, it may skid at the beginning and then roll. In such cases, the friction coefficient will change from dynamic to static according to the change in the particle behavior from skidding to rolling. Hence, the friction coefficient determined with the apparatus described above is to be the mean value during contact duration. So, the measured value should be called average friction coefficient.

Critical Impact Velocity for Spherical Particle: The relationship between the impact velocity and the average friction coefficient at the impact of steel shot with carbon steel surface are shown in Fig. (2-39) for various impact angle and particle diameter. In the range of lower impact velocity, it decreased with increasing velocity. This is because it is, as described above, the average of the dynamic friction coefficient during particle skidding at the beginning of contact and the static friction coefficient during particle rolling after stopping of skidding. In the range of higher impact velocity, it exhibited nearly constant level. This must be because the solid particle skidded for whole duration it contacted with the surface. Hence, the constant level of average friction coefficient independent of impact velocity must be that of the dynamic friction coefficient. As it is possible to distinguish the border line in two ranges so clearly that the velocity at which it reaches the level was, just for the convenience of measurement, newly defined as the critical impact velocity, though at impact velocities lower than but close to it small extent of skidding and accordingly cutting process may be involved.

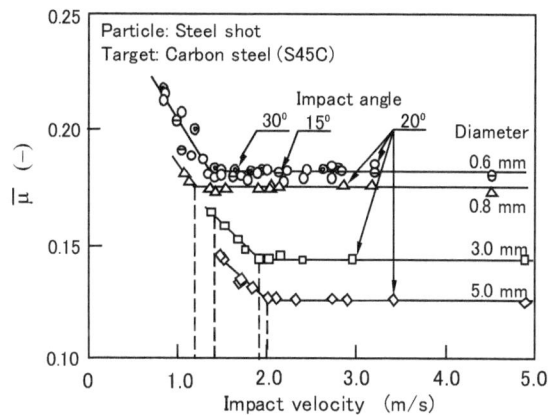

Fig. (2-39). Relationship between average friction coefficient and impact velocity for various impact angles and for various sizes of steel shot; target material, carbon steel [14].

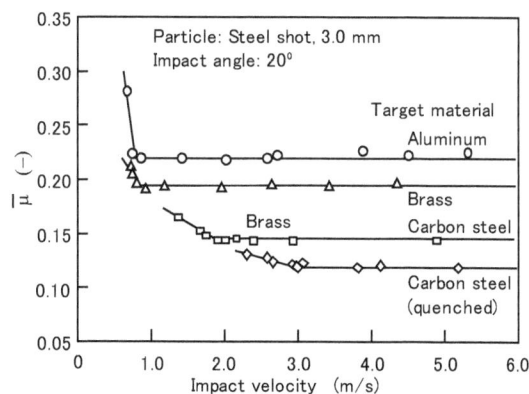

Fig. (2-40). Relationship between average friction coefficient and impact velocity for target specimens of various materials: particle, steel shot of 3.0 mm in diameter; impact angle, 20° [14].

It should be remarked in the figure that average friction coefficient is constant for the same size of steel sphere independent of the impact angel, and that with increasing particle diameter the friction coefficient decreased but the critical impact velocity increased. The results of measurements with a steel shot of 3.0 mm in diameter impinged on various target materials are shown in Fig. (**2-40**). It appears that the higher the target material hardness, the higher the critical impact velocity is.

Critical Impact Velocity for Angular Particle: The friction coefficient was measured for the impact of steel grid of 0.88 mm in mean diameter on carbon steel target. Some scattering of the friction coefficient was observed even under the same impact condition, so that frequency distributions of friction coefficient at four different impact velocities were measured which are shown in Fig. (**2-41**). The distributions at the impact velocities higher than 2.6 m sec^{-1} were similar to each other, but at 1.9 m sec^{-1} it was spread wider and the friction coefficient at the maximum frequency rose.

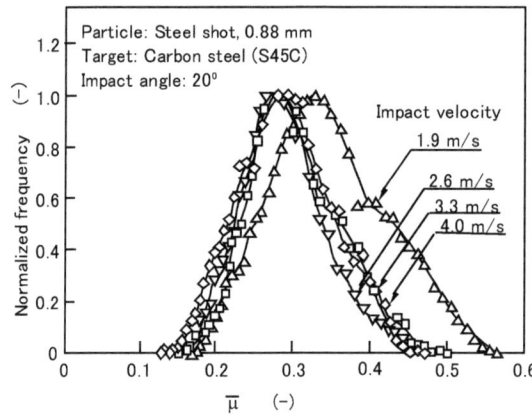

Fig. (2-41). Normalized frequency distribution of average friction coefficient during impact of angular particle for various impact velocities: particle, steel grid of 0.88 mm in mean diameter; target material, carbon steel; impact angle, 20° [14].

This result may be interpreted as follows: in the higher impact velocity range, all the particles must skid over the target surface resulting in an almost fixed level of friction coefficient, since it is dynamic one, so that the scattering might be attributed to the irregularity in the shape of the particles which caused the repellence in arbitral direction. In the lower velocity range, some of particles would but others would not skid at the target surface because of the irregular shape of particle, which would give rise not only in the level of friction coefficient but also in the degree of scattering of data.

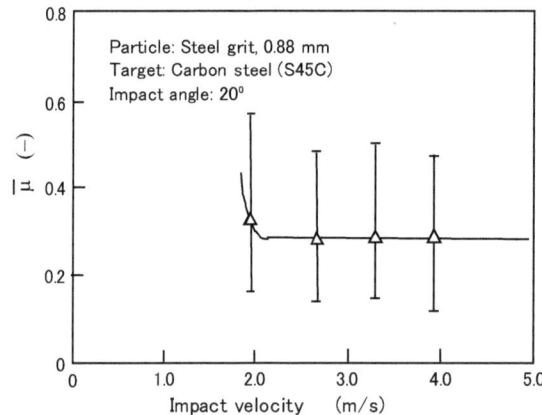

Fig. (2-42). Determination of critical impact velocity for angular particle: particle, steel grid of 0.88 mm in mean diameter; target material, carbon steel; impact angle, 20° [14].

The above interpretation leads to the adoption of the coefficient at the maximum frequency as the representative level at that impact velocity. The points obtained in the above mentioned procedure were connected with a line in the similar manner as the case of spherical particle to give the critical impact velocity of 2.2 m sec^{-1} as shown in Fig. (**2-42**). The measurements with silicon carbide and silica sand particle of various sizes on carbon steel target specimen showed that critical impact velocity of angular particle decreases with the increase in particle size.

Prediction of Critical Impact Velocity: The theoretical value of criticality impact velocity was deduced by Yabuki and the collaborators [15]. They simulated the skidding *vs.* rolling process of spherical particle impingement on target plane with the rolling *vs.* slip phenomenon of two circular cylinders which are rotating in different direction with different rotational speed with their axes paralleled. At the moment of contact of the cylinders, there is no slip. When the rotation of the cylinder advances to some extent, slippage would be generated in the tangential direction. But the attached surfaces actually stick to each other by the friction force for a while, during which the slippage is absorbed by the deformation of the subsurface materials. When the extent of the deformation exceeds a certain threshold extent, the slippage actually occurs. This process must correspond to that of the critical impact velocity: at the moment of oblique impingement of particle on material surface, compressive stress is generated in the material surface in the front of the particle and tensile stress at the rear side as shown in Fig. (**2-43**). The higher the impact velocity, the higher the tensile stress would be. It was assumed that the tensile strain reached at a certain threshold extent, the particle would slip, which must corresponds to that at the critical impact velocity.

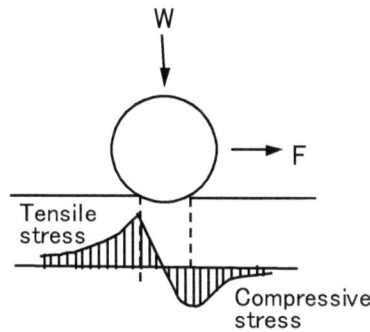

Fig. (2-43). Compressive stress and tensile stress generated in target material during oblique impact of solid particle [15].

$$V_C = \frac{K + cos\psi(cos\psi - \bar{\mu}sin\psi)}{\bar{\mu}(K + sin\psi) - cos\psi}$$

$$\times \frac{k_1 d}{(1+e)\left(\pi\sqrt{\dfrac{m}{2\pi p r_s}} - k_e V n^{-1/5}\right)}$$

$$\times \frac{Hv_s}{2.7\,E sin\,\theta}$$

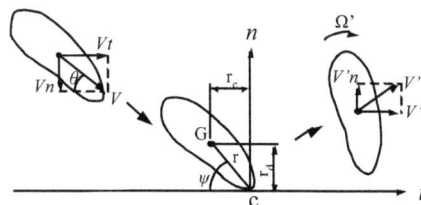

Fig. (2-44). Derivation of critical impact velocity: *m*, mass of particle; *I*, Moment of inertia; *E*, Young modulus; *e*, Restitution coefficient; *K*, shape factor of particle; *p*, yield pressure during impact; r_s, radius of solid particle corresponding to the sphere.

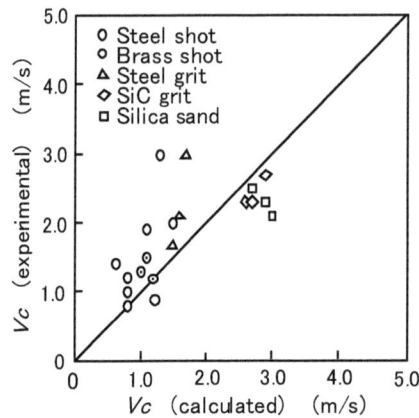

Fig. (2-45). Comparison between measured and calculated critical impact velocity [15].

With additional assumptions and equations to determine the forces which act on the impinging particle, spherical and angular, an equation was obtained for prediction of critical impact velocity for the combinations of various particles and target materials (Fig. (**2-44**)).

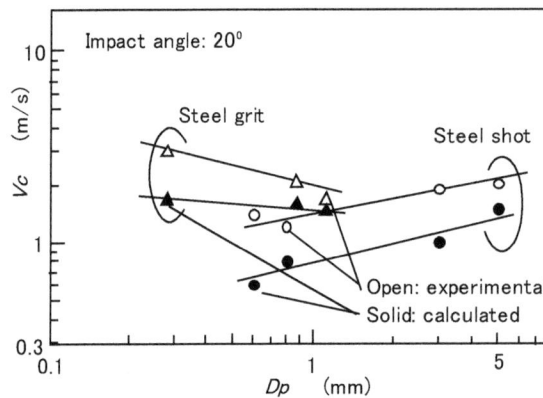

Fig. (2-46). Comparison in dependency to particle size between measured and calculated critical impact velocity [15].

The comparison between the calculated and measured critical impact velocity is given in Fig. (**2-45**). You can see the equation estimates the velocities low. Similar tendency may be recognized in the estimation of the dependence of the velocity on the particle size (Fig. (**2-46**)). It should be, however, also recognized that the equation estimated correctly the particle size dependence of the velocity which is reversed in the cases of spherical and angular particle.

Being connected with the description in the preceding sections, critical impact velocity was estimated with the equation for the impact of the gypsum particle on the pump impeller of the high chromium steel, which resulted in 17 m sec^{-1}. The average circumferential velocity of the impeller was nearly 18 m sec^{-1}. The relative velocity between slurry and impeller must be actually lower than this. In short, it can be sufficiently guessed that the damage was being generated by the particle impingement of under critical impact velocity. It was accordingly quite reasonable that the laboratory test result agreed with the performance in the real pump not only in the ranking order but also in the relative erosion rate.

2.6. Procedure for Evaluating Erosion-Proof Performance

The evaluation of slurry erosion-proofness of various materials may be carried out according to the algorithm in Fig. (**2-47**) as follows: firstly, the critical impact velocity is to be calculated with the properties of the solid particle and the candidate material for the actual machine; secondly, solid particle impact velocity is to be estimated basing on the operating conditions of the machine.

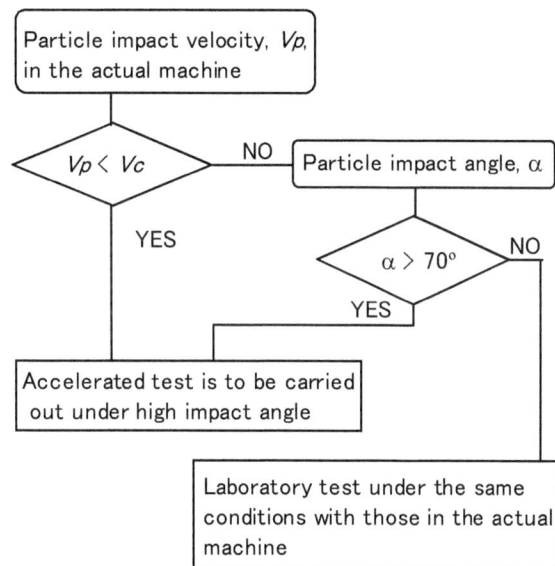

Fig. (2-47). Procedure for evaluating slurry erosion-proof performance of materials.

If it is lower than the critical impact velocity calculated in advance as described above, accelerated test is to be carried out under high impact angle. Even if the particle impact velocity is higher than the critical impact velocity, the same test as above may be carried out when the impact angle in the real machine is over 70°. If the slurry of real machine is used, such precise data which can predict the relative damage rate may be obtained.

3. TRUE NATURE OF EROSION

3.1. Erosion by Liquid Impact

According to the definition of technical term in Chapter 1, progressive loss of original material from a surface due to impinging liquid particle is classified in erosion together with those due to cavitation and solid particle impact. It is some times called in short as liquid impact erosion or rain erosion. The erosion of this mode occurs in the front edge of wings of aircraft which flies in the rain at high speed, in the elbow of pipeline in which two phase fluid, vapor and liquid, is transported with high velocity as well as in the rotor blade of steam turbine which recovers the power from the process steam of comparatively low-pressure. The remarkable feature of liquid impact erosion is to be more troublesome than the erosion of other type.

It must be not less troublesome to conduct material testing of liquid impact erosion in laboratory because it is not easy to prepare uniform sized and uniform shaped droplets in great number, to accelerate them up to a high velocity close to that of sound, and to let them impact on specimen surface repeatedly. At this point, material testing of cavitation erosion is much easier due to the vibratory testing apparatus. Then, how about diverting its testing purpose from cavitation erosion to rain erosion? In fact, in 1979 the comparison of damage mechanism between cavitation erosion and rain erosion was discussed to reach the view: commonly in both sort of erosion, localized but intensive impact load is repeatedly imposed to the material surface; rather small difference is in the distribution of impact intensity, wide spread in cavitation (Fig. (1-3)) as compared with narrow and sharp in rain erosion, since it is well established that the intensity of the impulsive pressure occurred at the droplet impact does not depend on the droplet size. Furthermore, the impact velocity in the erosion of turbine blades and aircraft wings must be rather uniform because it is that velocity of the blades and wings. Taking influence of other factors into consideration the following reserved conclusion was given: the ranking order of merit for candidate materials obtained in vibratory testing apparatus might coincide with that of anti-rain erosion performance in the field [16]. There was, however, no reliable evidence to support this conclusion because there is no reliable date for the ranking orders of merit in the performance of materials in the field. It is rather difficult in the field even to operate a machine keeping the conditions constant for a long duration of time. It is almost impossible to build similar machines of various materials one after another and operate them under the same conditions.

In 1987, the key to above problem was given by Staniša and his collaborators [17]. For several years, at every periodical maintenance, they completely dismantled a steam turbine to measure the surface roughness of the blade, and obtained a relationship between the mean depth of erosion damage, Y, and the mass of water impinged on blade, m' (Fig. (**2-48**)). You can recognize it on the graph that the tangents of linear relations converge to two points, Y_S, on the vertical line at $m' = 0$.

At nearly same time but independently to this, Oka and the collaborators found the linear relationship between the damage depth, d, and testing duration as well as the characteristic depth, D_C, in the cavitation erosion test with vibratory apparatus as described in the section at the beginning of this chapter. These points, Y_S and D_C, certainly support the possibility, from the testing results obtained in the vibratory cavitation facility, to predict not only the ranking order of performance materials but also relative damage rate in liquid impact erosion which materials would suffer in steam turbine and pipe lines in the field.

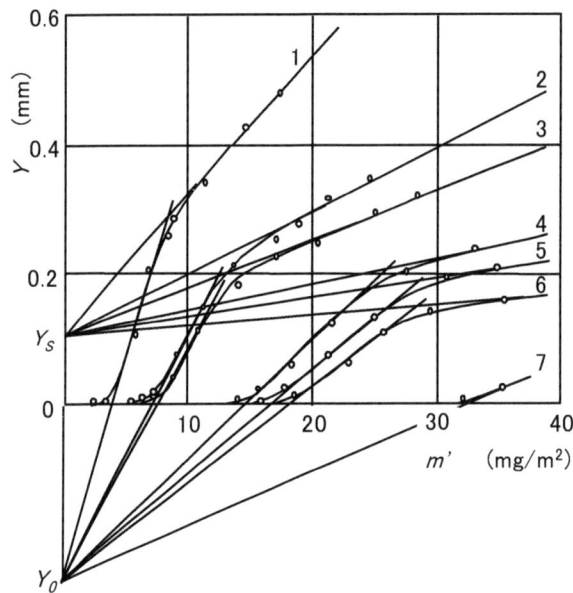

Fig. (2-48). Dependence of mean depth of erosion damage of turbine blade steel on the mass of water per unit surface area at various impinging velocities [17].

3.2. Plastic Deformation as Basic Mechanism for Erosion

In the preceding sections on liquid impact erosion, the convincing proof, that is, Y_S and D_C, was given for the hypothesis that the generation mechanism of liquid impact erosion may be the same with that of cavitation erosion. It had also been demonstrated in preceding sections that D_C represents the maximum extent of plastic deformation that can be accumulated during the incubation period before the initiation of material separation from the surface. It had been further made clear that even after the incubation period the erosion damage is controlled by the rate of plastic deformation generation.

As to the solid particle impact erosion, the nature of damage is strongly dependent on the characteristic surfaces in the crater, and accordingly on the way in which plastic deformation is generated.

So that, it may be concluded that the mechanism for generation of pure erosion damage in ductile metallic material, cavitation or solid and liquid impact it may be, is symbolized by plastic deformation in common.

REFERENCES

[1]   Hobbs JM, Bachman D. Environmentally controlled cavitation test. Characterization and Determination of Erosion Resistance ASTM STP 474. Philadelphia, WB: American Society for Testing and Materials 1970; pp. 29-47.

[2]   Louis H. Erosive Zerstörungen durch Strömungskavitation. Doktor-Ingenieur genehmige Dissertation, Technische Universität Hannover, 1973.

[3]   Matsumura M, Oka Y, Ueda M, Yabuki A. Prediction of service life of materials exposed to cavitation attack. Boshoku-Gijutsu (presently Zairyo- to-Kankyo) 1990; 39: 550-5.

[4]   Sakamoto A, Isomoto Y, Matsumura M. Index of cavitation damage to metallic materials. Zairyo-to- Kankyo 1994; 43: 76-81.

[5]   Yabuki A, Noishiki K, Matsumura M. A method for predicting the damage rate of cavitation erosion in actual Machines. Zairyo-to-Kankyo 2000; 49: 489-93.

[6]   Noishiki K, Yabuki A, Komori K, Matsumura M. A method for predicting the incubation period of cavitation erosion. Zairyo-to-Kankyo 2000; 49: 483-8.

[7]   Ozaki T, Onuma T. Development of a cavitation-erosion resistant stainless steels for sea water hydraulic machines. Boshoku-Gijutsu (presently Zairyo-to-Kankyo) 1987; 36: 83-90.

[8]   Stauffer WA. Wear of metals by sand erosion. Met. Prog. 1956; 69 [1]: 102-7.

[9]   Matsumura M, Oka Y, Hatanaka H, Yamawaki M. Erosion-corrosion testing method. Boshoku-Gijutsu (presently Zairyo-to-Kankyo) 1980; 29: 65-9.

[10] Sheldon GL. Similarities and differences in the erosion behavior of materials. J. of Basic Engineering, Trans. ASME, Series D 1970; 92: 619-26.

[11] Hojo H, Tsuda K, Cao MT. Erosion damage of poly-methyl-methacrylate. J. Soc. Mater. Sci. 1980; No.332: 731-35.

[12] Matsumura M, Oka Y, Yamawaki M. In: slurry erosion-corrosion of commercially pure iron in fountain-jet testing apparatus. Proc. 7th Int. Conf. on Erosion by Liquid and Solid Impact; 1987: Cambridge, UK: 40.

[13] Oka Y, Matsumura M, Yamawaki M, Sakai M. Jet-in-slit and vibratory methods for slurry erosion-corrosion tests of materials. In: Miller JE, Schmidt FE Jr., Eds. Slurry Erosion: Uses, Applications, and Test Methods; 1984: Denver USA: ASTM STP 946, Philadelphia, American Society for Testing and Materials, 1987; pp. 141-54.

[14] Yabuki A, Matsuwaki K, Matsumura M. Critical impact velocity in the solid particles impact erosion of metallic materials. Wear 1999; 233-235: 468-75.

[15] Yabuki A, Matsuwaki K, Matsumura M. Theoretical equation of the critical impact velocity in solid particles impact erosion. Zairyo-to-Kankyo 1998; 47: 631-7.

[16] Matsumura M. Liquid particle impact erosion and cavitation erosion. J. Mater. Sci. Soc. Jpn. 1979; 16: 57-64.

[17] Stanisa B, Dičko M, Puklavec K. Erosion process on the last stage rotor blades of turbines in service. In: Field JE, Dear JP, Eds. Proc. 7th Int. Conf. on Erosion by Liquid and Solid Impact; 1987: Cambridge, England: 16.

CHAPTER 3

Combined Erosion and Corrosion

Abstract: Pure erosion and pure corrosion were intentionally superimposed and their interaction was observed, expecting that some useful information on the mechanism of erosion-corrosion might be obtained. As a pure erosion process cavitation erosion and slurry erosion were chosen. It has already been made clear in the previous chapter that the damage generation processes of the erosion are both pure mechanical. As a result of experiments, it was found that cavitation erosion was at the initial stage accelerated by corrosion but in the later stage it was inhibited through the adsorption of chloride ions on the metal surface. Alternatively, the erosion always accelerated corrosion by renewing the metal surface. Slurry erosion process, that is, the damage to the metal surface by the impact of solid particle was entirely enhanced by corrosion. Corrosion of the target material was sometimes enhanced and sometimes inhibited depending on the particle impact angle, which was in good accordance with the observation in the previous chapter. The most important discovery in the experiments was that under the superposition of erosion and corrosion the product of the cooperation of erosion and corrosion did not exist. Accordingly erosion-corrosion was judged as a pure electrochemical corrosion process.

Keywords: Superimpose, cavitation erosion, corrosion, adsorption, chloride ion, ductile fracture, brittle fracture, slurry erosion, particle impact angle, erosion-corrosion.

1. SUPERPOSITION OF CAVITATION EROSION AND CORROSION

1.1. Introduction

It may be rather exaggerating to claim that erosion-corrosion is a product of the cooperation of erosion and corrosion. They must, however, in some ways concern in erosion-corrosion. In clarifying the mechanism of erosion-corrosion it may be one of the functional approaches to superimpose the processes of pure erosion and pure electrochemical corrosion and observe the interaction. In the previous chapter cavitation erosion was proved to be a pure mechanical process, so that it may be chosen as pure erosion. Superposition of cavitation erosion and corrosion implies the deterioration of a material due to the simultaneous action of a mechanical force caused by cavitation and aggressive attack of the environment. The components would probably assist each other to bring about a larger amount of damage than the simple sum of the damage caused by each component separately.

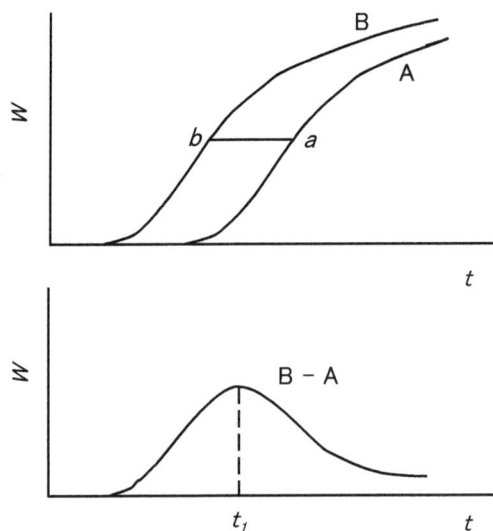

Fig. (3-1). Superposition of cavitation erosion and corrosion [1].

In the study of the superposition of the components, the total damage to a material can be easily measured. Difficulties arise, however, in determining the proportion of those due to each component to the total damage. There is no reliable method to determine the damage of each component separately under the superposition. One might expect that the difference between the total damage under the superposition and the sum of damage caused by each component may be an indication of an increase in damage resulting from the interaction of these components. This difference is, however, no other than a feign influence of the interaction because erosion rate as well as corrosion rate is time dependent. In order to explain this we assume that the cumulative weight loss, W, of specimen caused by cavitation attack can be expressed by curve A in Fig. (3-1), and that pure corrosion is negligibly small in amount but substantially reduces the duration of incubation period of the cavitation erosion process. Then, the cavitation erosion and corrosion process will become curve B in the figure.

The difference between curves A and B is shown in the figure as curve B-A, which was obtained by subtracting A from B at the identical testing duration. This curve implies that the influence of corrosion on the cavitation erosion reaches a maximum at the testing time of t_1, which is by no means related to the assumption that the incubation period has been reduced. It is, therefore, indispensable for determining the true corrosion influence to compare points which correspond to each other, for example, the points *a* and *b* on each curve in the figure. It is quite clear that the amount of damage is equal, but *b* occurs at a somewhat earlier time. This judgment exactly coincides with the above premise. Such mark points, however, are scarcely found on erosion curves. Instead, the author has found that a cavitation process consists of at least three different kinds of characteristic periods which were discerned not by the superficial rise and fall of the erosion rate curve but by the fracture mechanism of the material. These characteristic periods are expected to be used for the mark points in cavitation erosion process.

As to the amount of corrosion, we cannot neglect it even though it may further complicate the situation. So, it is postulated in this section that the total damage of a metal under the superposition of erosion and corrosion, Wt, consists of W' and F', where W' is the amount of erosion or the amount of material which separates itself from the surface as small metallic particles, and F' is the amount of corrosion which dissolves from the surface as metallic ions. Thus,

$$Wt = W' + F' \qquad\qquad\qquad\qquad\qquad (3\text{-}1)$$

The purpose of the description in this section is substantiation of the above basic equation, namely the concept of superposition. For this purpose, combined cavitation erosion and corrosion test was carried out on test specimens of commercial pure iron using the vibrating specimen facility [1].

1.2. Experimental Apparatus and Procedure

Cavitation was induced by a vibratory specimen facility shown in Fig. (3-2). The test specimen (column of $16^{\phi} \times 22$ mm) was vibrated axially at a high frequency of 19.9 kHz with the double amplitude of 24 μm in test liquid contained in a beaker. De-ionized water saturated with nitrogen gas (specific resistance of over 5 $\times 10^{6}$ Ω cm) was used as test liquid to cause pure cavitation erosion (W) on the test specimen. Environmental corrosive liquids were used to cause the superposition of the erosion and corrosion (Wt). To cause pure corrosion (F) on the test specimen, the corrosive test liquid was circulated through the container and the test specimen was held still in the test liquid without any vibration.

Aqueous solutions of hydrochloric acid and sodium hydroxide were chosen as corrosive environments. This is due to the fact that the corrosion behaviors of iron and steel in those solutions have been well elucidated. The temperature of test liquid was kept constant at 40 °C.

The material used for the investigation was commercial pure iron with the following chemical composition: C, 0.02; Si, 0.19; Mn, 0.25; P, 0.010; S, 0.004. The test specimen that had been machined out on the lathe received the following standard pre-treatment: grinding with emery paper, annealing in a vacuum and then buff-polishing. To measure the amount of corrosion, Fe^{2+} ion content in the test solution was analyzed by the colorimetric procedure, using a photoelectric filter photometer and o-phenanthroline hydrochloride reagent.

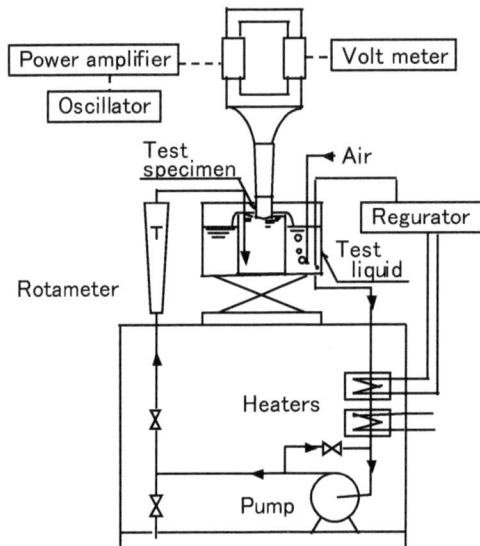

Fig.(3-2). Experimental apparatus used for superposition of cavitation erosion and corrosion [1].

1.3. Pure Erosion Process, *W*

As has already shown in Fig. (**2-2**) of the preceding chapter, in the latter half of weight loss period in vibratory erosion test many small holes with diameter smaller than 1 mm appear on the specimen surface. Fig. (**3-3**) shows a relationship between the total area of small holes, *a*, and testing time, *t*, for a commercial pure iron specimen (marked by solid circles, dimension in right hand side axis). The cumulative weight loss of the specimen (*W*) versus *t* curve is also shown in the diagram (marked by open circles, dimension in left hand side axis). The curve of *a vs. t* can be divided into four stages, in each of which the curve can be approximated by a straight line. During the first period **a** only the roughness of damaged surface increases but no holes appear. In **b** period a few small holes appear in the central region of the specimen, but the number of holes does not increase rapidly. In **c** period, the number of holes as well as the area of each hole increases rapidly and the whole surface of the specimen is nearly covered with holes. In the last period **d**, the generation as well as the growth of holes almost stopped.

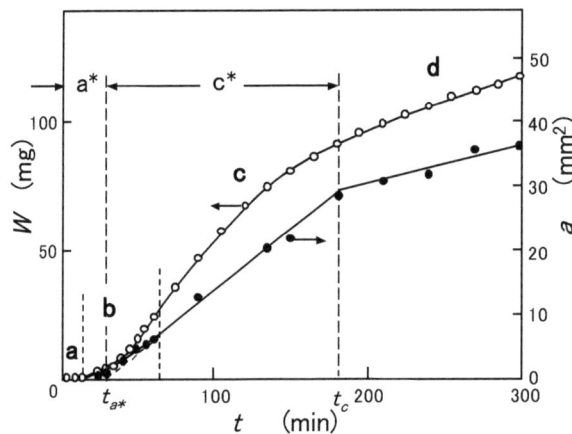

Fig. (3-3). Typical process of cavitation erosion represented with *W-t* and *a-t* curve [1].

The rate of erosion reaches a maximum in the **c** period where the number and the area of the holes increase rapidly, and it goes down in **d** period. It has been clarified that the damage rate in the **d** period decreases because of spare cavitation bubbles sitting in each deep hole on the damaged surface [2, 3]. Thus, the reduction in damage rate in the **d** period is caused by the change in the behavior of cavitation bubble on the specimen surface and independent of the properties of material. The **b** period is considered to be merely a transition period from **a** to **c**.

The surface of the test specimen was observed by SEM and it was found that on the surface damaged in the **a** period small fracture with plastic deformation overlapped each other in a complicated way, so that the fracture could be regarded as ductile. On the contrary, in the **c** period many smooth surfaces are found on the sidewall of holes. This indicates brittle fracture.

The above observations and considerations led us to define the characteristic periods a* and c*, the characteristic amount of damage W_{a*}, W_{c*} and W_{abc}, and the coefficient W_{c*}/W_{abc}: W_{abc} is the sum of damage in the periods **a**, **b** and **c**; W_{c*} is the amount of damage in c* period; c* period is the period between the time t_{a*} obtained by extending the *a-t* line and the finishing time of the **c** period; the coefficient W_{c*}/W_{abc} is the ratio of the amount of damage in c* period to the total amount of damage or the ratio of the contribution of brittle fracture to the total damage.

By the way, the time point when the break occurred on the *d-t* line in Fig. (**2-7**) must correspond to t_{a*} mentioned above.

1.4. Pure Corrosion Process, *F*

It can be easily found out in textbooks or handbooks that the corrosion rate of iron and steel depends on the pH of the environmental liquid. The effect of the pH of aerated water on the corrosion rate of mild steel is shown in Fig. (**3-4**). This was obtained by Whitman *et al.* using hydrochloric acid and sodium hydroxide to adjust pH [4].

On the same figure, our measurements on commercial pure iron are also plotted. Fairly good agreement is obtained with the results of Whitman *et al.* From this figure, the following three characteristic environments were chosen and the corrosion process in each environment was minutely investigated.

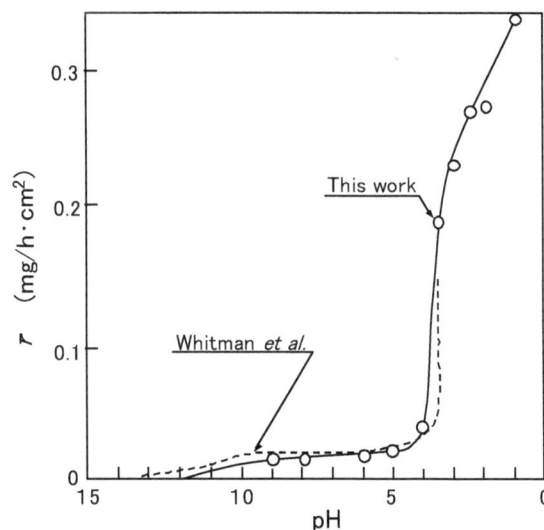

Fig. (3-4). Effect of pH of test liquid on corrosion rate of specimen of mild steel and pure iron [1].

Acid Region (Hydrochloric Acid Solution of pH 2): The relationship between the amount of corrosion, *F* (mg), and the testing time is shown in Fig. (**3-5**) by open circles. Since it was impossible to find smooth line passing through all the experimental points, three smooth curves were selected, each of which fitted the points well in their region.

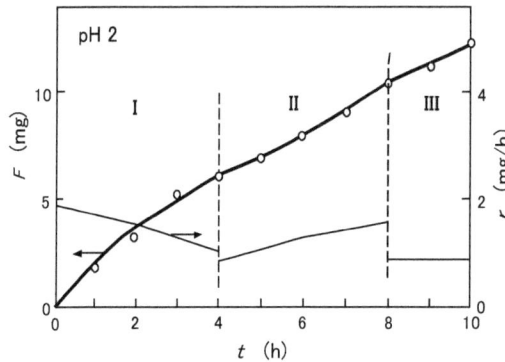

Fig. (3-5). Corrosion processes of test specimen in environmental liquid of pH 2 [1].

Consequently, the corrosion rate obtained by differentiating the F-t curve is expressed by three different lines. Although common experience indicates that the corrosion rate should change continuously with time, the discontinuity of the corrosion rate curves in this work clearly implies the existence of three different corrosion processes. Even though it is an expedient measure, this approach is quite useful for engineering application. These characteristic corrosion periods are to be recognized in the figure. In the period I, the corrosion rate decreases monotonically from the beginning of the test until the testing time of 4 hours; then it decreases in period II, and in period III it remains constant.

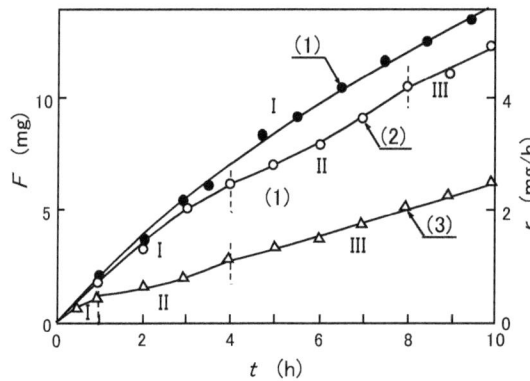

Fig. (3-6). Effect of pre-treatment of test specimen on corrosion behavior; (1) machining followed by polishing, (2) machining followed by annealing followed by polishing, (3) machining followed by annealing [1].

In order to elucidate the corrosion mechanism in each period, a series of corrosion tests was conducted using several test specimens with different types of pre-treatment. The results obtained are shown in Fig. (3-6) in which curve (2) was obtained from the specimen which received pre-treatment in the standard order, that is, machining followed by annealing and then polishing. The test specimen without annealing, curve (1), showed a longer duration of period I and the specimen without polishing, curve (3), showed a shorter duration of period I than the standard specimen (2). The appearance of the specimen surface in each characteristic period was also observed by SEM.

Based upon the experimental results and the above observations, the corrosion mechanisms in each characteristic period can be explained as follows:

Period I: This period was extended by the omission of annealing and reduced by the omission of polishing. This clearly implies that the surface layer of the specimen which was strained by machining and polishing dissolves within this period. Whitman *et al.* stated that in acid conditions the surface film of ferrous hydroxide was neutralized and dissolved to bring the metal surface into direct contact with the main solution, thus reducing the resistance to mass transfer; evolution of hydrogen gas and reduction of oxygen took place simultaneously; the former caused turbulence at the metal surface and thereby increased the rate of oxygen diffusion; thus dissolution of metal was proceeded at high rate.

Hydrogen evolution was also observed in this experiment. This must not only be due to the total acidity in the main solution as Whitman *et al.* have pointed out, but also due to some reduction of hydrogen overvoltage caused by the activation of the metal surface through straining, because the evolution of hydrogen gas decreased gradually with time and ceased completely at the end of period I. This is evidently due to the dissolving of the strained layer followed by the exposure of bulk metal without any strain in it to the liquid. Thus, the rate of corrosion continues to decrease from the beginning of the test through period I as the strained layer dissolved away.

Period II: It could be seen in the SEM observation that there were many tiny steps on the corroding surface in this period, which had a certain directional conformity to the crystalline grain [1]. This is evidently the so-called "hill and valley " structure which is well known to form on dissolving crystalline materials owing to the solubility difference among directions. The existence of this structure on the corroding surface proves that in this period the surfaces of the bulk metal without any strain dissolve with different rates depend on the crystallographic directions. The crystallographic planes with a higher dissolving rate preferably dissolve away and at the last stage of the period only the planes with the lowest dissolving rates remain on the specimen surface.

Period III: The specimen surface is almost covered with crystallographic planes of least dissolving rate. The dissolutions in this period are, therefore, limited to grain boundaries and neighboring impurities which are more active than a bulk of crystal. The corrosion rate is reduced in this period because such active spots are limited in number on the corroding surface.

Neutral Region: The *F-t* curve for a specimen which was held still in a hydrochloric acid solution of pH 4 and the *r-t* curve obtained by differentiating the *F-t* curve are shown in Fig. (3-7). The amount of corrosion for a testing duration of 10 h is as small as one tenth of that which occurs at a pH of 2. The corrosion rate decreases monotonically from the beginning of the test. No evolution of hydrogen gas was observed.

Whitman *et al.* stated that in this neutral region the surface of corroding steel was covered with a thin film of ferrous hydroxides which protected the steel from corrosion by reducing the rate of oxygen diffusion; the film required a considerable time to develop its full protectiveness. The gradual reduction of the corrosion rate shown in the figure is in accordance with this explanation. Even after 10 h of testing no such characteristic period appeared as in the case of the acid region.

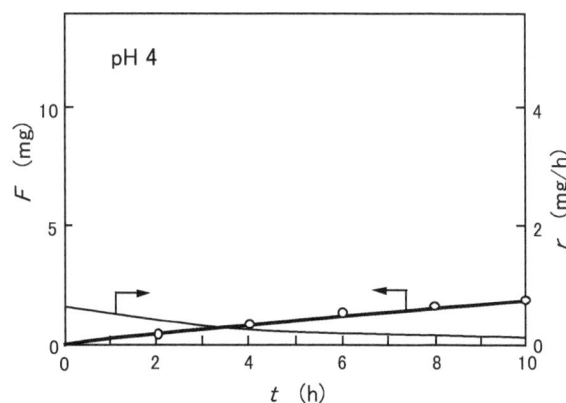

Fig. (3-7). Corrosion process of test specimen in environmental liquid of pH 4 [1].

Alkaline Region: A test specimen was held still in a sodium hydroxide solution of pH 12. After a test of 10 h duration, no ferrous ion was detected in the solution by the colorimetric analysis. Consequently the corrosion rate of the specimen in this region confirmed to be zero. This must be evidently due to the rapid formation of a strong protecting film on the specimen surface just as was pointed out by Whitman *et al.*

1.5. Total Damage under the Superposition of Erosion and Corrosion, *Wt*

Fig. (**3-8(a)**) shows the results of the combined cavitation erosion and corrosion tests, that is, cumulative weight loss, *Wt*, of test specimen exposed to cavitation in environmental liquids. The curve of pure cavitation erosion (*W*) is also shown as a dotted line in the figure, that is, the cumulative weight loss of the test specimen exposed to cavitation in the de-ionized water saturated with nitrogen gas (D.W.).

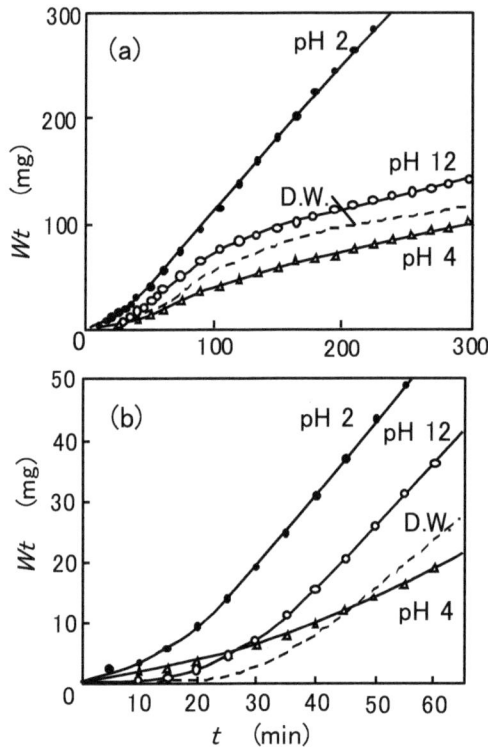

Fig. (3-8). Cumulative weight loss, *Wt,* of test specimen exposed to cavitation in environmental liquids: (a) overview; (b) initial progress [1].

The early stage of the *Wt-t* relation is magnified in Fig. (**3-8(b)**), where the order in the amount of damage is pH 2 > pH 4 > pH 12 ≅ D.W. Conversely, for the last stage (at the testing time of 300 min, for example) the order is pH 2 > pH 12 > D.W. > pH 4. It is noteworthy that the order in the amount of damage is changed with the lapse of testing time.

Table 3-1. Average rate of damage in a* and c* period [1]

	t_{a*}	t_c	$t_c - t_{a*}$	W_{abc}	W_{a*}	W_{c*}	W_{c*}/W_{abc}	W_{a*}/t_{a*}	$W_{c*}/(t_c - t_{a*})$
	(min)	(min)	(min)	(mg)	(mg)	(mg)	(-)	(mg/h)	(mg/h)
D.W.	30	180	150	90.4	2.8	87.6	0.696	5.60	35.04
pH 2	94	180	86	223	98	125	0.561	62.55	87.21
pH 4	25	66	41	22.5	5.0	17.5	0.778	12.00	25.61
pH 12	16	66	44	36.8	1.2	35.6	0.967	4.50	48.35

What attracts special interest is that the amount of damage in pH 4 in the last stage is smaller than that of de-ionized water. The suspicion that it might be due to scattering in the measurements was dispelled, because all repeated runs conducted under the same conditions gave the same order.

As each of the *Wt-t* curves in Fig. (**3-8**) bears a resemblance to that of pure erosion (the curve in Fig. (**3-8**) labeled as D.W. or the *W-t* curve in Fig. (**3-3**)), their *a-t* relation was examined. As a result a* and c* periods were clearly obtained for all of the *Wt-t* curves. Thus the *Wt-t* curves can be compared with the *W-t* curve with sufficient physical significance as they have common features. In Table **3-1**, values of t_{a*}, t_{c*}, W_{a*} and W_{abc} are listed, where W_{a*}/t_{a*} and $W_{c*}/(t_c-t_{a*})$ are averaged damage rates in the a* and c* periods, respectively. As regards the a* period, the damage rate in the liquid of pH 2 increased as much as eleven times over that of pure erosion, 2.5 time in pH 4 and no change in pH 12. For the c* period, however, the damage rate increased by 2.5 times in pH 2, 1.4 times in pH 12 and it decreased to 0.8 times in pH 4.

1.6. Mutual Interaction of Erosion and Corrosion

Dissolution Rate under Combined Erosion and Corrosion Condition, F': It is impossible to compare directly the combined erosion and corrosion process with that of the pure corrosion because no common feature has been recognized between them. In order to estimate the dissolution rate under a combined erosion and corrosion condition, therefore, the following indirect method was adopted. The specimen was exposed to cavitation for certain duration in the middle of a corrosion test and the change in the corrosion rate induced by the exposure was examined.

Dissolution rate in acid region: A test specimen was removed from the testing bath in the middle of a corrosion test to be exposed to cavitation for 1 h in the de-ionized water saturated with nitrogen gas and then it was put back into the testing bath to continue the corrosion test. Fig. (**3-9**) shows the experimental results in which the exposure to cavitation was given, (a) in the characteristic period II (t = 6 h), and (b) in the period III (*t* = 8 h). The relationship between the amount of corrosion (*F*) and the testing time is given by open circles, but the weight loss of the specimens during the exposure to cavitation is not given in the figure.

Exposure to cavitation produced a sudden increase in the corrosion rate regardless of the characteristic period (solid lines in the figures). The increase in the corrosion rate was larger than that at the beginning of the test but what is worthy to remark is the change of corrosion rate with testing time after the exposure. It decreased monotonically with a striking resemblance to the change of corrosion rate in the early stage of period I.

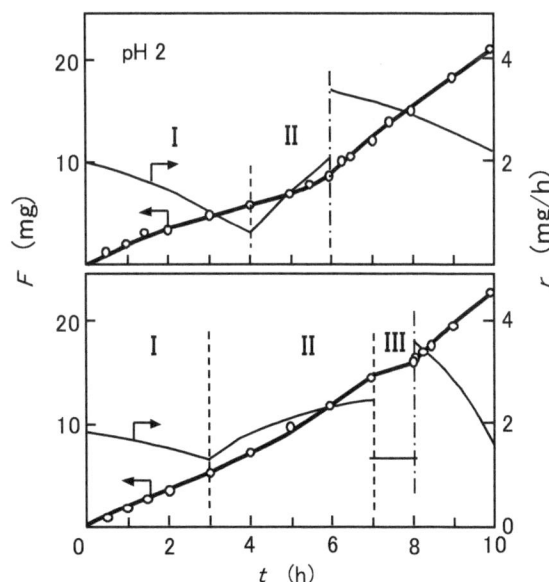

Fig. (3-9). Responses of corrosion rate (pH 2) to cavitation attack applied for 1 h in II period (upper) and in III period (lower) [1].

These results prove that cavitation attack produced the same effects on the specimen surface as machining and polishing, raising the corrosion rate at the beginning of the test.

Dissolution rate in neutral region: 1 h exposure to cavitation after a 4 h corrosion test produced a sudden increase in corrosion rate.

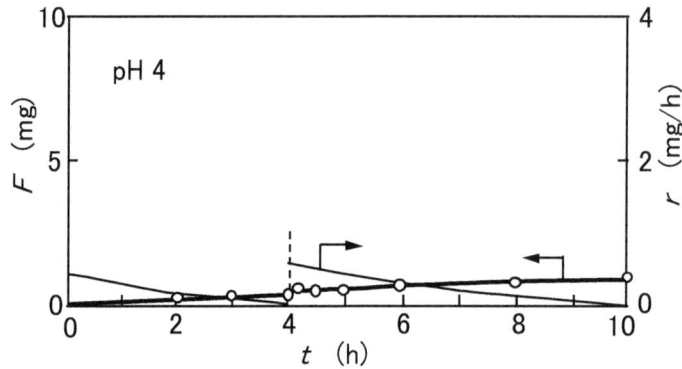

Fig. (3-10). Response of corrosion rate (pH 4) to cavitation attack applied for 1 h [1].

The gain as well as the gentle decrease in corrosion rate thereafter bears a striking resemblance to the early stage of the corrosion test, Fig. (**3-10**). It is easily interpreted that cavitation attack has destroyed the protective film over the specimen surface and brought it to the same state as at the beginning of the test.

Dissolution rate in alkaline region: No ferrous ion was detected in the testing liquid of pH 12 even after one hour exposure of the test specimen to cavitation. The corrosion rate, therefore, was confirmed to be zero as well. Though the protective film over the surface was destroyed by the cavitation attack, dissolution was interrupted possibly by rapid recovery of the film. This was the same situation as occurred at the beginning of the corrosion test in the liquid of pH 12.

In conclusion, the effects of cavitation attack on corrosion may be described as follows. The cavitation attack not only removes the protective film on the surface but also produces strain in the surface layer of the specimen in the same way as machining and polishing do. These actions of cavitation will keep the dissolution rate the same as that at the beginning of the corrosion test. The amount of dissolution during the cavitation attack of t h in an environmental liquid will be given by the following equation with enough accuracy for engineering use.

$$F' = r_0 t \qquad\qquad (3-2)$$

Here, r_0 is the corrosion rate at the beginning of the corrosion test in an environmental liquid.

According to the equation, the amount of dissolution under superposition of erosion and corrosion in any liquid is no less than that of pure corrosion, which in turn indicates that the reduction of cavitation erosion corrosion damage in the liquid of pH 4 should be attributed to some reduction of the metal particle separation rate under the superposition.

Separation Rate under Combined Erosion and Corrosion Condition, W':

Separation rate in a* period: Table 3-1 shows that the average damage rates during a* period in the liquid of pH 2 and pH 4 are larger than that of pure erosion, and no significant increase occurs in the liquid of pH 12. The effects of environment on the separation rate were investigated using a similar experimental technique to that described in the previous section: a test specimen was removed from the testing bath in the middle of a cavitation test in the de-ionized water with nitrogen gas, dipped into an environmental corrosive liquid and held there for 1 h and then put back into the testing bath to continue the cavitation test.

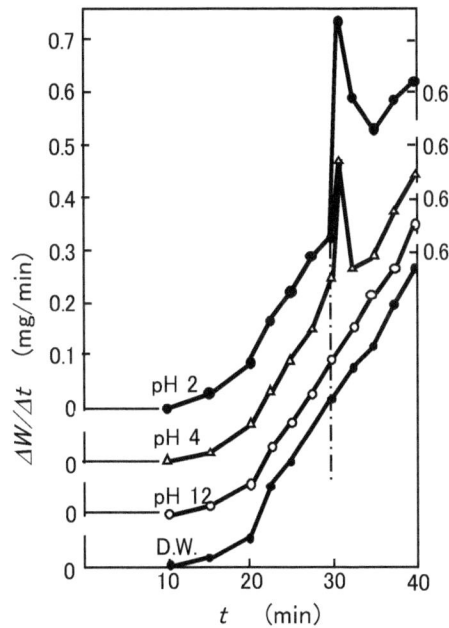

Fig. (3-11). Response of $\Delta W/\Delta t$ in period a* to dipping in various environmental liquids; duration of dip, 1 h [1].

The "difference erosion rate" $\Delta W/\Delta t$ before and after the dipping is shown in Fig. (**3-11**), here, ΔW is the weight loss of the specimen in the testing duration of Δt. About 2 mg of dissolution was observed during the dipping but it is not shown in the figure. The dipping of the specimen into the liquid of pH 2 and pH 4 produced an evident increase in the erosion rate. No significant change, however, was seen when it was dipped into the liquid of pH 12 or into the de-ionized water which was adopted for the blank test.

Results of an experiment under the same conditions but with a dipping duration of 5 min are shown in Fig. (**3-12**), in which no significant change was produced in the erosion rate by any environment liquid.

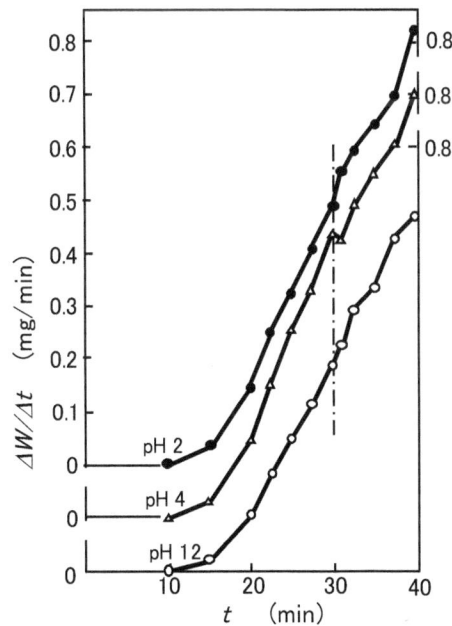

Fig. (3-12). Response of $\Delta W/\Delta t$ in period a* to dipping into various environmental liquids; duration of dip, 5 min [1].

In both experiments, the interruption of the pure erosion test was conducted at 30 min after the beginning of the test. Time is sufficient within a* period and the resulting weight loss is enough for measurement accuracy.

Separation rate in c* period: Table **3-1** shows that the average damage rates during c* period in the environmental liquid of pH 2 and pH 12 are larger than that in the de-ionized water (pure erosion). Conversely in the environmental liquid of pH 4, it is lower than that of pure erosion.

A specimen was removed from the de-ionized water bath in the middle of the cavitation erosion test (t = 150 min, c* period) and dipped into an environmental liquid and held there for 1 h. Then it was put back to the bath to continue the cavitation erosion test. The changes in the difference rate $\Delta W/\Delta t$ caused by dipping are shown in Fig. (**3-13**).

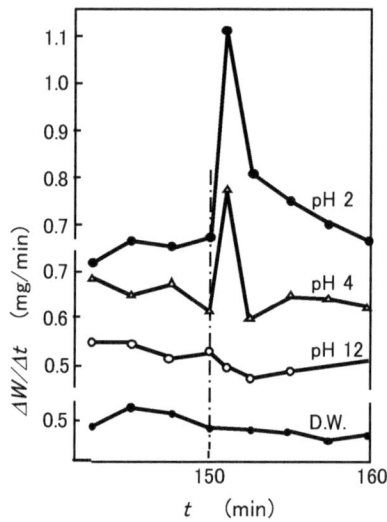

Fig. (3-13). Response of $\Delta W/\Delta t$ in period c* to dipping into various environmental liquids; duration of dip, 1 h [1].

An increase in the rate was observed at pH 2 and pH 4. These results, however, did not agree with those of Table **3-1**. Fig. (**3-14**) shows the results of another experiment in which the duration of dipping was decreased to 5 min. An evident decrease occurred in the damage rate by the liquid of pH 4, but no significant change was brought about by the liquid of pH 2 and pH 12, which again does not agree with the results in Table **3-1**.

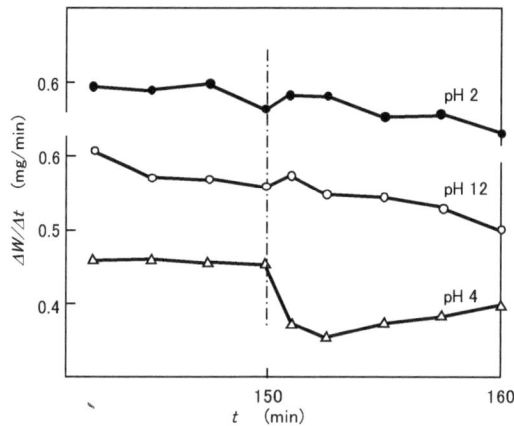

Fig. (3-14). Response of $\Delta W/\Delta t$ in period c* to dipping into various environmental liquids; duration of dip, 5 min [1].

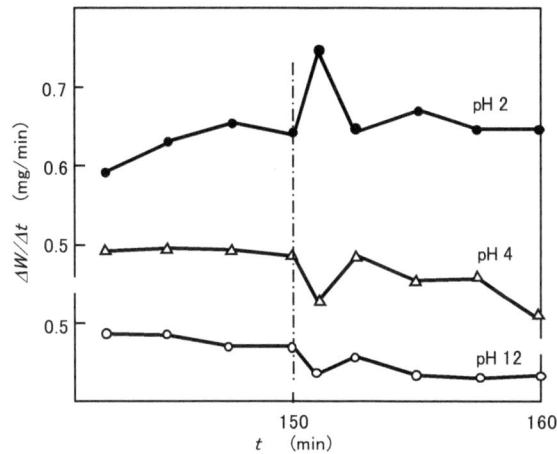

Fig. (3-15). Response of $\Delta W/\Delta t$ in period c* to dipping into various environmental liquids; duration of dip, 15 min [1].

Fig. **(3-15)** shows the results of 15 min dipping, where the changes of the rate caused by the liquid of pH 2 and pH 4 are inversed to each other. Although a small change in rate is observed in the liquid of pH 12, it may be concluded that the results in Fig. **(3-15)** are in agreement with those of Table **3-1**.

The experimental results above suggest strongly some participation of chloride ion (Cl⁻) adsorption because the decrease in erosion rate was brought about by the short duration dipping only. With the aim of exploring the possibility of adsorption, an experiment was carried out in the following way: a test specimen was removed from the de-ionized water bath in the middle of an erosion test ($t = 150$ min) and dipped into an environmental liquid for 5 min. It was then dipped into a nitric acid solution (pH 4) and a voltage of 15 V was applied to it for 15 min in order to delete the adsorbed Cl⁻ ions from the surface. Then it was put back in the de-ionized water bath to continue the cavitation test. The experimental results for each environmental liquid are shown in Fig. **(3-16)**, where the curve labeled as B.T. is the result from a specimen which was not dipped into any environmental liquid but into the nitric acid solution with a voltage of 15 V for 15 min.

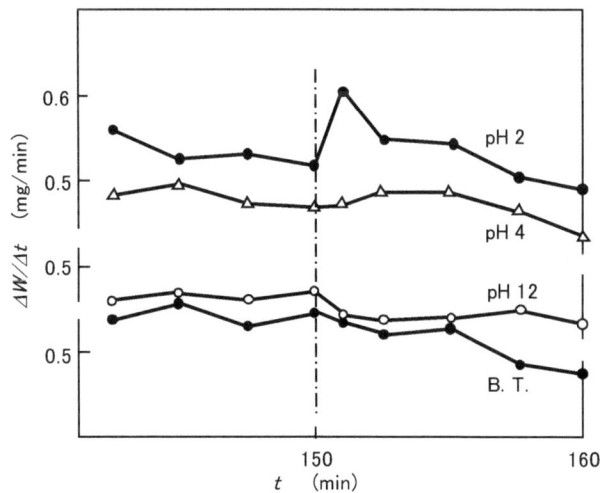

Fig. (3-16). Response of $\Delta W/\Delta t$ to desorption of Cl⁻ ion [1].

The curve proves that the desorption treatment, that is, the treatment to delete the adsorbed ions does not affect the erosion rate so long as no Cl⁻ ion is adsorbed on the surface. It is clearly seen that an apparent increase appeared in the erosion rate for pH 2 and the decrease in the rate for pH 4 disappeared (cf. the results shown in Fig. (**3-14**) where the duration dipping into the environmental liquids was also 5 min but no desorption treatment followed).

The appearance of the damaged surface (t = 150 min) just before and after the dipping into each environmental liquid for 5 min was observed by SEM to find the development of hair cracks and increase of the crack width in the case of pH 2. In the case of pH 4 and pH 12, however, no change was seen even at the place where some changes would be expected if it were exposed to the liquid of pH 2 [1].

Based on the above experimental results, the effects of the environmental liquid on the cavitation damage in c* period can be summarized as follows: the environments exert two opposing kinds of influence on the erosion rate; one is to accelerate erosion by extending and widening the cracks on the damaged surface, another is to inhibit the erosion through the adsorption of Cl⁻ ion on the surface. In the environmental liquid of pH 2, the influence of the former is stronger than the latter and the damage is larger than that for pure erosion. In the liquid of pH 4 the former is weaker than the latter and the damage is less than that for pure erosion. In the liquid of pH 12 both influences are too weak to be detected by any change in $\Delta W/\Delta t$, but the priority of the first effect is proved by the Wt-t curve (Fig. (**3-8**)) that is larger than the pure erosion (D.W.). The inhibitory influence of environments is not seen in the damage of a* period. This fact in turn explains how adsorbed Cl⁻ ion inhibits the damage: it is not through releasing the surface from the impulsive pressure of cavitation bubble but through inhibiting the development of hair cracks on the damaged surface which aid brittle fracture in c* period.

Estimation of the Separation Rate under the Combined Erosion and Corrosion Condition, W': As a clear difference in the influences of environment on the separation rate has been noticed between a* and c* periods, the amount of separation under the combined erosion and corrosion condition should be estimated for each characteristic period. Hence within the range of this experiment,

$$W' = f_{a*}R_{a*}t_{a*} + f_{c*}R_{c*}(t_c - t_{a*}) + f_d R_d(t - t_c) \qquad (3\text{-}3)$$

where R_{a*}, R_{c*} and R_d is an average erosion rate in a*, c* and **d** period respectively; f_{a*}, f_{c*} and f_d are coefficient which represent the intensity of environmental influence. t_{a*} and t_c should be found under the combined erosion and corrosion condition. Each term on the right side of the equation is the amount of separation in each characteristic period.

1.7. Estimation of Total Damage under the Combined Erosion and Corrosion

The total damage caused to a specimen under the combined erosion and corrosion condition can be obtained by substituting Eqs. (3-2) and (3-3) for W' and F' in Eq. (3-1), thus,

$$Wt = f_{a*}R_{a*}t_{a*} + f_{c*}R_{c*}(t_c - t_{a*}) + f_d R_d(t - t_c) + r_o t \qquad (3\text{-}4)$$

The **d** period may be excluded because the change of R in this period is caused by the change in the behavior of cavitation bubble on the specimen surface and is not related to any characteristics of the material. The following equation ignores the decrease of the erosion rate in the **d** period. Consequently it may overestimate the total damage in this period. Nevertheless this estimation will be accurate enough for engineering use.

$$Wt = f_{a*}R_{a*}t_{a*} + f_{c*}R_{c*}(t - t_{a*}) + r_o t \qquad (3\text{-}5)$$

In Fig. (**3-17**) the total damage obtained in this experiment is compared with that calculated from Eq. (3-5), where r_0 was obtained from Figs (**3-5**) and (**3-7**), t_{a*} and t_c from Table **3-1**, and f_{a*} and f_{c*} were calculated from the data in Table **3-1**. The overestimations in early stage are attributed to the negligence of the incubation period and those in the late stage are to the negligence of **d** period as discussed in the previous section. The overestimations are always on the safe side for design purposes.

Thus, the basic equation, Eq. (3-1), or the concept of superposition of erosion and corrosion has been substantiated valid for the processes of cavitation erosion and corrosion.

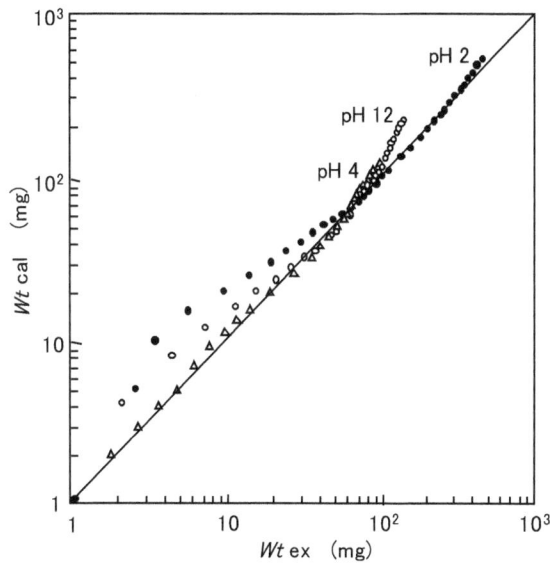

Fig. (3-17). Comparison between calculated and experimental value of *Wt* [1].

2. SUPERPOSITION OF SLURRY EROSION AND CORROSION

2.1. Background

Slurry erosion is advantageous in superposition test of erosion and corrosion at next two points. Firstly, it has been established that the solid particle impact erosion is a pure mechanical process more ensured than it is for cavitation erosion. Secondly, the effect of testing time does not need to be considered because the rate of slurry erosion does not change with the time. On the other hand, the effect of the solid particle impact on the corrosion rate of target metal surface drastically changes with the impact angle, as it was described in the previous chapter. So that, the jet-in-slit apparatus was improved for the impact angle to be able to be accurately adjusted.

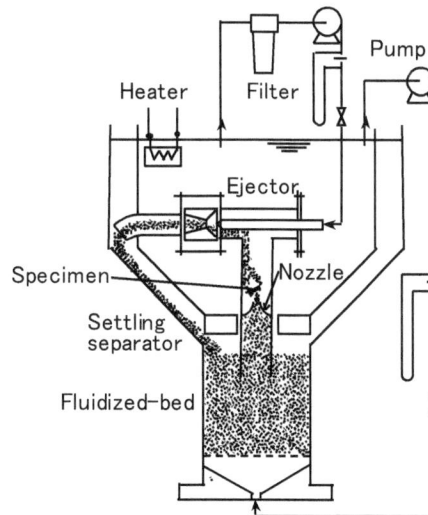

Fig. (3-18). Schematic drawing of fountain-jet testing facility [5].

The schematic diagram of fountain-jet apparatus, that is, improved jet-in-slit apparatus is shown in Fig. (3-18). An ejector was newly installed at the top of the test section to suck up the slurry from the fluidized bed below. The stream of slurry of particle concentration of 60 % was accelerated by the reducer nozzle and allowed to impinge on the test specimen at a velocity of 3 m sec^{-1} [5].

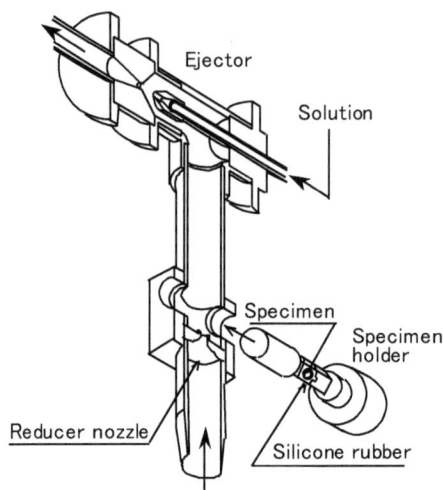

Fig. (3-19). Details of test section [5].

In order to restrict the test surface to a small area where the impact conditions uniform the test specimen, a cylinder 5 mm in diameter and 5 mm in height, was held between two matching sections of silicon rubber which were held together on a bolt. Only the top flat surface was exposed to the impinging slurry since the bottom as well as the side wall of the specimen was entirely covered by vinyl tape. The impact angel of the slurry stream on the specimen surface was regulated by turning and then clamping the support bolt. As a result, it was established that the impact angle of solid particle was equal to the angle between the slurry stream and the surface of the specimen (Fig. (**3-19**)). The technique used for the measurement of particle impact conditions in this facility have been reported elsewhere [6], and only the results of the measurements are listed in Table **3-2**.

The test specimen was of commercial pure iron as well as the cavitation erosion test. The test liquid was de-ionized water pH being adjusted with hydrochloric acid solution and caustic alkali of sodium solution.

Table 3-2. Solid particle impact conditions in fountain-jet facility [5]

Impact velocity	Impact angle	Impact frequency
(m/s)	(deg)	(1/cm^2 s)
3.0	10-80	ca. 10^5

2.2 Experimental Results

Pure Erosion Process: Pure erosion tests on commercial pure iron were conducted using the slurry of silica sand and de-ionized water.

The results obtained are shown in Fig. (**3-20**) where the relationship between erosion rate and particle impact angle (α) is given. The erosion rate in mg cm^{-2} min^{-1} (solid line in the graph) was divided by the flow rate of impacting particles to obtain an alternative erosion rate expressed in mg kg^{-1} (broken line). The shift of the peak of the curve results from the fact that the flow rate of impacting particles depends on the angle, α, between the specimen surface and the slurry stream. In this chapter the erosion rate expressed in mg cm^{-2} min^{-1} is adopted since corrosion rate is expressed in this unit as well.

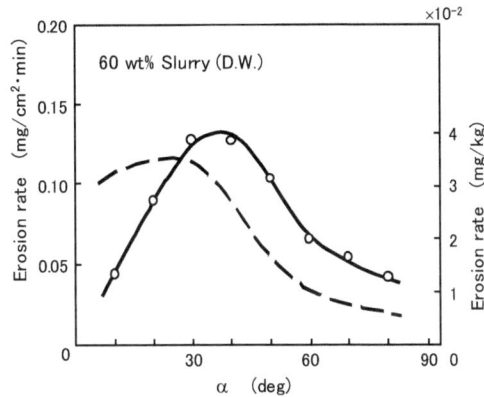

Fig. (3-20). Effect of impact angle on erosion rate for commercial pure iron specimen in fountain-jet facility: jet velocity, 3.0 m sec^{-1}; slurry, de-ionized water and silica sand of 60 wt%, 40 °C [5].

Pure Corrosion Process: Pure corrosion tests were conducted by circulating aqueous solutions without solid particles. As shown in Table **3-3** the average corrosion rate for an hour period was found to depend on the pH of the test solution but to be independent of the angle at which the specimen was set. A comparison between Table **3-2** and Fig. (**3-3**) makes it clear that corrosion rates and erosion rates were of the same order of magnitude except for the corrosion rate at pH 12.

Table 3-3. Corrosion rate of commercial pure iron specimen in fountain-jet facility [5]

Corrosiveness	Corrosion rate (mg/cm^{2}min)
pH 2	0.07
pH 4	0.025
pH 12	0

Combined Erosion and Corrosion: Combined erosion and corrosion tests were carried out in slurries of silica sand and corrosive solutions. The results are shown in Fig.(**3-21**) through (**3-23**). The thick solid line in each graph represents the combined erosion and corrosion rate, $\dot{W}t$, as a function of particle impact angle. The broken line represents the pure erosion rate, $\dot{W}$, the dotted line the pure corrosion rate, $\dot{F}$, and the dashed line the sum of the pure erosion and pure corrosion rate, that is $\dot{W}+\dot{F}$.

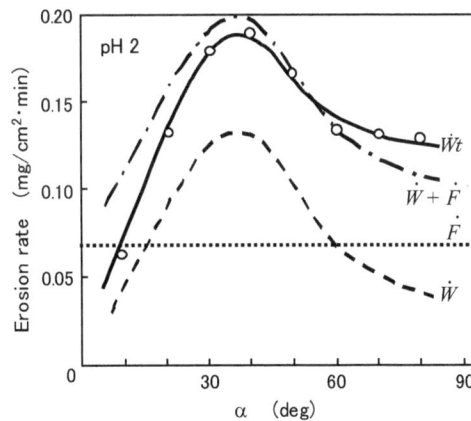

Fig. (3-21). Effect of impact angle on damage rate for commercial pure iron specimen; jet velocity, 3.0 m sec^{-1}; slurry, HCl acid solution of pH 2 and silica sand of 60 wt%, 40 °C [5].

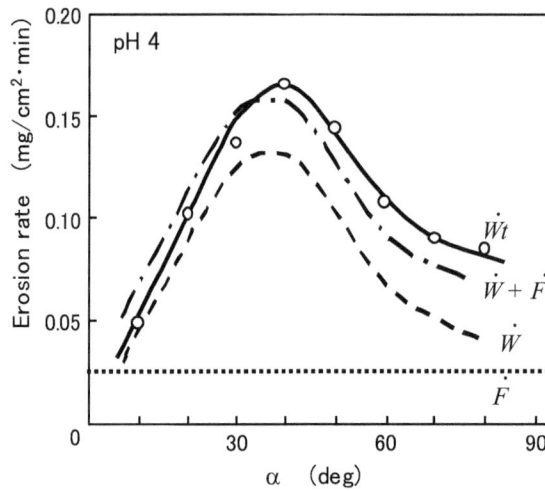

Fig. (3-22). Effect of impact angle on damage rate for commercial pure iron specimen; jet velocity, 3.0 m sec^{-1}; slurry, HCl acid solution of pH 4 and silica sand of 60 wt%, 40 °C [5].

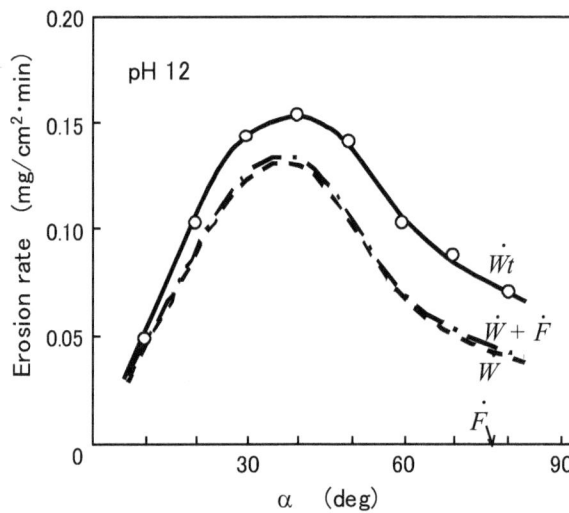

Fig. (3-23). Effect of impact angle on damage rate for commercial pure iron specimen; jet velocity, 3.0 m sec^{-1}; slurry, HCl acid solution of pH 12 and silica sand of 60 wt%, 40 °C [5].

In the slurry of pH 12 (Fig. (**3-23**)), the sum of the erosion and corrosion rate ($\dot{W} + \dot{F}$) is essentially identical with the erosion rate ($\dot{W}$); this results from the fact that the pure corrosion rate ($\dot{F}$) is essentially zero for all impact angles.

Nevertheless, the combined erosion and corrosion rate ($\dot{W}t$) is higher than the sum of pure erosion and pure corrosion rate. Thus some mutual acceleration of the rates of erosion and corrosion must be occurring. In other words, the sign of the mutual interaction is positive.

In the slurry of pH 4 (Fig. (**3-22**)), $\dot{W}t$ is still higher than $\dot{W} + \dot{F}$ at high impact angles. At low impact angles, however, a change in behavior occurs and thereafter $\dot{W}t$ is consistently lower than $\dot{W} + \dot{F}$. This is interpreted to mean that the erosion and corrosion rates are inhibited when they occur together, or that their mutual interaction is inhibitory. The same situation is observed in the slurry of pH 2 (Fig. (**3-21**)), but the change in behavior occurs at a higher angle.

2.3. Synergy of Erosion and Corrosion

As described in the previous chapter the damage rate of metallic material in solid particle impact erosion was not dependent on the time. In this experiment it did not change with time in pure erosion test as well as in the superposition of erosion and corrosion. In addition to this pure corrosion rates were substantially constant for the testing duration of one hour. Therefore, in the case of superposition of slurry erosion and corrosion, unlike the case of the cavitation erosion we may determine the amount of synergistic effect by simply subtracting the sum of the amount of pure erosion and pure corrosion from the total damage to a material, namely with the following equation.

$$Wt = W + F + W_S \tag{3-6}$$

Here, W_S is the amount of synergistic effect. This may be differentiated to obtain similar equation for damage rates:

$$\dot{W}t = \dot{W} + \dot{F} + \dot{W}_S \tag{3-7}$$

The extent of synergy obtained by subtracting $(\dot{W} + \dot{F})$ from $\dot{W}t$ is given in Fig.(3-24) as a function of impact angle with the parameter of pH of environmental liquid.

Surprisingly, several results of completely contrary to the expectation were included: the negative synergy at shallower impact angles in the slurry of pH 2 and pH 4. So far, negative synergistic effects have been reported in literatures which were written by researchers in Japan, China, England as well as Norway using similar equation to Eq. (3-6). And, all the researchers attributed the negative synergy, in fact, to the scattering of measurements.

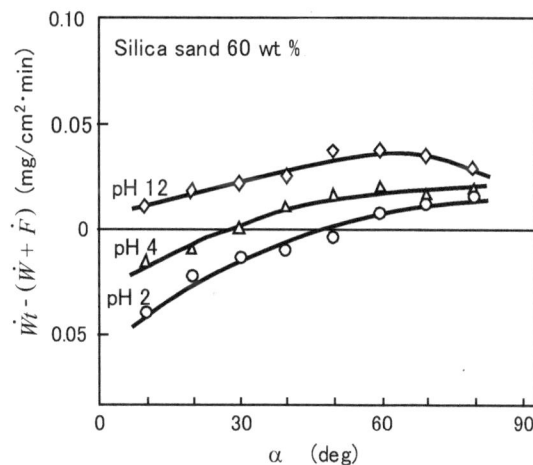

Fig. (3-24). Effect of impact angel on the amount of synergistic effect of erosion and corrosion for commercial pure iron specimen; jet velocity, 3.0 m sec^{-1}; slurry, corrosive liquids and silica sand of 60 wt%, 40 °C [5].

Those in Fig. (**3-24**), however, cannot be attributed to the scattering of measurements because it can be read that it continuously changes from the positive value by gradually lowering to the negative value. Furthermore, the negative value was most in the slurry of pH 2 in which the corrosion rate must be highest. Thus, it is clear that the results do not make sense at all, and we cannot trust the equation which brought about such meaningless result. We had to confirm the reliability of the equation in advance before we used it. In order to prove this equation we have to obtain independently each term of the equation and confirm that the extent of the left-hand side term agrees with the sum of those of right-hand side. However, it is impossible to obtain the synergy term independently. Thus, the proof of Eq. (3-6) is not completed.

2.4. Estimation of Total Damage under Combined Erosion and Corrosion

For the interpretation of the observed results, application of the basic equation, Eq. (3-1), which was previously developed for combined cavitation erosion and corrosion was here again attempted. The equation postulates that the total damage to a metal under the superimpose of erosion and corrosion, that is, Wt consists of W' and F' where W' is the amount of erosion, that is, the amount of material which separates itself from the surface as small metallic particles, and F' is the amount of corrosion or the amount of material which dissolves from the surface as metallic ions. Differentiation of Eq. (3-1) gives:

$$\dot{W}t = \dot{W}' + \dot{F}' \tag{3-8}$$

Each term of the equation has dimension of rate, namely, mg cm^{-2} min^{-1}. In order to prove the validity of the equation, each term was determined independently and quantitatively by using the following graphic methodology:

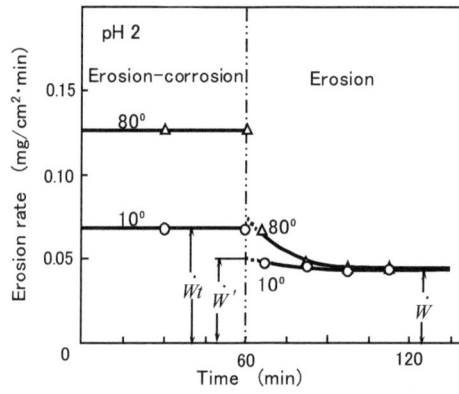

Fig. (3-25). Combined erosion and corrosion test followed by pure erosion test on commercial pure iron specimens; jet velocity, 3.0 m sec^{-1}; slurry, HCl acid solution of pH 2 and silica sand of 60 wt%, 40 °C [5].

As shown in Fig. (3-25) combined erosion and corrosion test which was conducted in the corrosive slurry of pH 2 was suddenly, at a testing time of 60 min, switched over to pure erosion test by exchanging the corrosive slurry with a non-corrosive one, namely slurry of silica sand and de-ionized water. The change in damage rate at particle impact angles of 10° and 80° which was brought about by the exchange of the slurries is shown in the figure: The damage rate decreased gradually to eventually reach the individual pure erosion rate (compare with Fig. (3-20)). The extrapolation of the curve back to the time of the slurry exchange was considered to give the erosion rate under the combined erosion and corrosion condition, namely $\dot{W}'$. The same procedure was adopted to obtain $\dot{F}'$, but in this case the combined erosion and corrosion tests were followed by pure corrosion tests circulating only clear liquid of pH 2 (Fig. (3-26)).

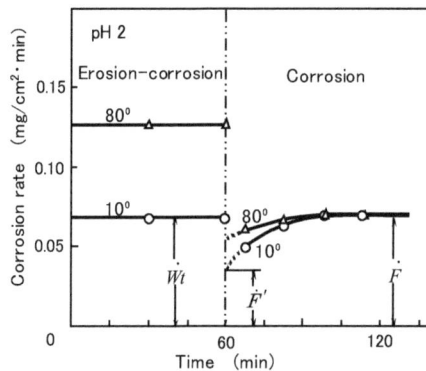

Fig. (3-26). Combined erosion and corrosion test followed by pure corrosion test on commercial pure iron specimens in environmental liquid of pH 2; jet velocity, 3.0 m sec^{-1}; temperature of liquids; 40 °C [5].

The values of $\dot{W}'$ and $\dot{F}'$ thus obtained for impact angle of 10° and 80° are listed in Table **3-4**. Excellent agreement is obtained between the observed value of $\dot{W}t$ and the sum of $\dot{W}'$ and $\dot{F}'$.

Table 3-4. Comparison of damage rates [5]

Impact angle	$\dot{W}'$	$\dot{F}'$	$\dot{W}' + \dot{F}'$	$\dot{W}t$
(deg)	(mg/cm²·min)			
10	0.047	0.031	0.078	0.067
80	0.075	0.053	0.128	0.125

As a reason for higher $\dot{W}'$ than $\dot{W}$ the rise of the surface roughness and dissolution of the strain hardened layer in the surface was considered and it was verified by experiments [5]. The experimental result that $\dot{F}'$ was lower than $\dot{F}$ agreed well with that of Fig. (**2-30**) in the previous chapter.

The results shown in Figs. (**3-25**) and (**3-26**), and including Table **3-4** indicate that the extrapolation method described above is useful for determining each component under the combined erosion and corrosion condition, and that Eq. (3-1) is considered to be valid for slurry erosion and corrosion of commercial pure iron in the slurry of silica sand and corrosive solution.

3. TRUE NATURE OF EROSION-CORROSION

The purpose of this chapter is, as described in the introduction, superposition of pure erosion and pure corrosion to observe the resulting damage behavior with the aim to find some clue to mechanism of erosion-corrosion. The results so far obtained are as follows. The following equation for material balance is valid in both superposition of cavitation erosion and corrosion as well as slurry erosion and corrosion.

$$Wt = W' + F' \tag{3-1}$$

Here, Wt is the total damage to the material, W' the erosion damage under the influence of corrosion, F' the corrosion damage under the influence of erosion. Alternatively, the following equation with synergy term (W_S) was found very unacceptable for the superposition of slurry erosion and corrosion in this experiment in spite of the fact that it has been commonly used for synergy of pure erosion (W) and pure corrosion (F).

$$Wt = W + F + W_S \tag{3-6}$$

Based on above experimental results, together with the feasible premise that erosion-corrosion process does not include erosion process, that is, plastic deformation of the material we can select the true colors of erosion-corrosion from the terms above by method of elimination: W_S is first eliminated as it proved no sense at all; W as well as W' is eliminated according to the premise; F' is eliminated because there is no influence of erosion as there exists no erosion itself; what remained is F. Accordingly, the true nature of erosion-corrosion is pure electrochemical corrosion, which is the conclusion obtained in this chapter.

REFERENCES

[1]   Oka Y, Matsumura M. Cavitation Erosion-Corrosion. In: Field JE, Corney NS, Eds. Proceedings 6th International Conference on Erosion by Liquid and Solid Impact; 1983: Cambridge, England: Cavendish Lab. 1983; paper 11.

[2]   Suezawa Y, Matsumura M, Nakajima M, Tsuda K. Studies on Cavitation Erosion. J. of Basic Engineering, Trans. ASME, Ser. D 1972; 94: 521-531.

[3]   Plesset MS, Devine RE. Effect of Exposure Time on Cavitation Damage. J. of Basic Engineering, Trans. ASME, Series D 1966; 88: 691-705.

[4]   Whitman GW, Russell RP, Altieri VJ. Effect of Hydrogen-Iron Concentration on Submerged Corrosion of Steel. Ind. Eng. Chem. 1924; 16: 665-670.

[5]   Matsumura M, Oka Y, Yamawaki M. Slurry Erosion-Corrosion of Commercially Pure Iron in Fountain-Jet Testing Facility. In: Field JE, Dear JP, Eds. Proceedings 7th International Conference on Erosion by Liquid and Solid Impact; 1987: Cambridge, England: Cavendish Lab. 1987; paper 40.

[6]   Oka Y, Matsumura M, Ohsako Y, Yamawaki M. Particle Impact Conditions in Vibratory Sand-Erosion Facilities. Boshoku-Gijutsu (presently Zairyo-to-Kankyo) 1984; 33: 278-283.

CHAPTER 4

Theory of Electrochemical Corrosion

Abstract: Three kinds of impinging jet, namely free jet, submerged jet and jet-in-slit were introduced for producing erosion-corrosion on the specimens of copper and copper alloys. The complete process of impingement attack was reproduced with submerged jet and jet-in-slit but not with free jet: the origin of this sort of erosion-corrosion is the separation of protective oxide layer from the metal surface due to shear force as well as turbulence force; free jet caused only the shear force. When the flow direction of test liquid was reversed in jet-in-slit the turbulence in the flow disappeared, and instead, characteristic flow velocity distribution or fixed vortex was produced on the specimen surface. In accordance with those the localized corrosion with the morphology similar to the differential flow-velocity corrosion or the horseshoe corrosion appeared. The theory of macro-cell corrosion renders a comprehensible rationale to the relationship between the morphology of the localized corrosion and the characteristic flow pattern of liquid as follows: a difference in the flow condition on a metal surface causes the difference in the anodic dissolution rate of the metal, which induces the formation of macro-cell of corrosion. Once a macro-cell is formed the corrosion rate, in particular, the metal dissolution rate at the macro-anode is accelerated through "macro-cell current effect" as well as "surface area ratio effect". The measurement of macro-cell current was actually carried out during the progress of erosion-corrosion on a jet-in-slit specimen, and a clear difference was recognized in the behavior of anodic polarization curves at the corresponding locations. Thus, it was demonstrated that the erosion-corrosion on the copper alloy is electrochemical, localized corrosion.

Keywords: Free jet, submerged jet, jet-in-slit, impingement attack, shear force, turbulent force, fixed vortex, uniform corrosion, localized corrosion, micro-cell corrosion, macro-cell corrosion, polarization curve.

1. REPRODUCTION OF EROSION-CORROSION IN LABORATORY

1.1. Introduction

Syrett's theory of erosion·corrosion described in Chapter 1, that is, the separation of oxide layer due to the shear stress of the flowing fluid is simple, and easy to be accepted because the oxide layers on some copper alloys are soft and coarse just like cotton. Nevertheless some questions remain unanswered: how the oxide layer is separated in horseshoe shape; why the deposit attack occurs downstream from the barnacles but not at the ceiling immediate overhead of them where the shear stress is seemingly higher. According to the calculation of hydrodynamics, the shear stress on the inside wall surface of heat exchanger pipe monotonously decreases from the inlet end to the outlet end. Then, why does the damage occur at the distance of one or one and half pipe diameter downstream from the inlet end but not at the top end where the shear stress is at maximum? The differential flow-velocity corrosion of centrifugal seawater pump of gray cast iron in which corrosion damage was mild at the periphery of the spiral part, but it was severe near the shaft hole, cannot be attributed to the shear stress or the mass transfer enhancement in flowing fluid since these issues increase with the increase in flow velocity.

As to the wall thinning of carbon steel pipe in pure water at raised temperatures there is a serious questions on the FAC model: how it can explain the characteristic temperature dependency of wall thinning rate, that is, the sharp maximum, 4 mm y^{-1}, at near 140 °C; why this sort of damage sometime occurs and does not even under the apparently same operation conditions. The model cannot predict how the environmental parameters such as dissolved oxygen concentration and pH of liquid exert influence on damage rate. The engineers in the power stations are therefore in trouble lacking in certain foundation for the choice of operating conditions as well as the selection of material for boilers.

There is no room of discussion for the fact that all those problems clamed above come from the fact that the mechanism of the phenomenon has not sufficiently been clarified yet. Even though it is not yet clarified, however, if there is a testing facility with the reliability, or in which erosion-corrosion always and surely arises, it must be useful for obtaining the information on the effect of various parameters, and the decision of material selection and operating condition will be easier. Further, it will be largely helpful to the elucidation of the erosion-corrosion mechanism.

1.2. Jet-in-Slit

Test Apparatus and Test Results: During the combined erosion and corrosion study described in the previous chapter a unique testing facility was found by the chance for pure corrosion test: a clear corrosive solution was allowed to flow through the stationary specimen cavitation facility without any vibration in expectation to originate pure corrosion on the specimen. This was eventually a sort of impinging jet, injecting the solution in the slit between test specimen and the nozzle, and in fact apparent pure corrosion resulted, even though it was not uniform, but localized corrosion accompanied with particular damage pattern on the surface. The conclusion in the previous chapter was that the true nature of erosion-corrosion is pure electrochemical corrosion, but it was not accompanied by the additional condition whether it is uniform or localized. Thus, the test result got by this testing method seemingly regulated more clearly the definition of erosion-corrosion.

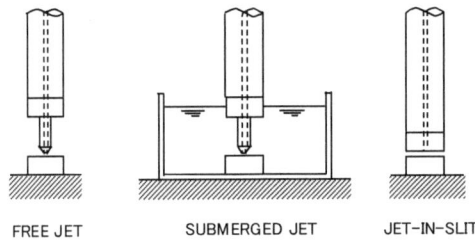

Fig. (4-1). Three types of impinging jet used in laboratory experiment [1].

Two other types of impinging jet were included in experiment in addition to the one mentioned above, namely free jet and submerged jet. Free jet is, as illustrated in Fig. (**4-1**), the same as the one used in conventional impingement tests in which the liquid jet strikes at right angles with the specimen surface in the air. Submerged jet is a jet submerged in water.

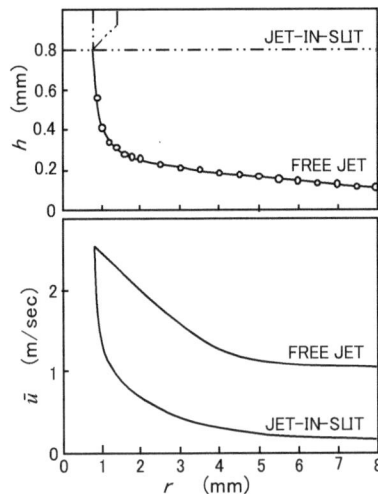

Fig. (4-2). Liquid film depth distribution on free jet specimen (upper) and radial flow velocity distribution on specimens of free jet and jet-in-slit (lower) [1].

In the jet-in-slit, jet is injected into the narrow gap. The jets contain the geometry and the flow conditions of the test liquid in common: the inside diameter of the nozzle is 1. 6 mm; the gap between the nozzle top end and the specimen surface is 0. 8 mm; and the flow rate of test liquid is 0. 4 L min^{-1}; at this flow rate, the fluid velocity at the nozzle outlet was 3. 3 m sec^{-1} and the Reynolds number at that point was 8100. The liquid film depth distribution under the above conditions on the free jet specimen is shown in Fig. (**4-2**) upper. The distributions of average radial flow velocity which were obtained by dividing flow rate with the apparent cross section of flow are also shown in the figure (lower). The flow velocity of free jet is higher than that of jet-in-slit because the liquid film of free jet is thinner than that of jet-in-slit in which liquid flow fills up the gap. The flow velocity distribution of submerged jet is presumed to be intermediate between the free jet and the jet-in-slit.

A test with 6/4 yellow brass specimen (16 mm in diameter with a chemical composition of 59. 77 copper, 0. 28 iron, and the remainder zinc) in 3% sodium chloride solution (40 °C) indicated that the weight loss of the specimens after 24 h was 9. 7, 10. 0, and 11. 3 mg for the free jet, submerged jet, and jet-in-slit, respectively. Inspection of surface and cross section of the specimen after the tests showed that extreme corrosion was seen with submerged jet and jet-in-slit as shown in Fig. (**4-3**). The cross section was measured with a surface roughness meter, so that the longitudinal direction is magnified 100 times with respect to the transverse direction.

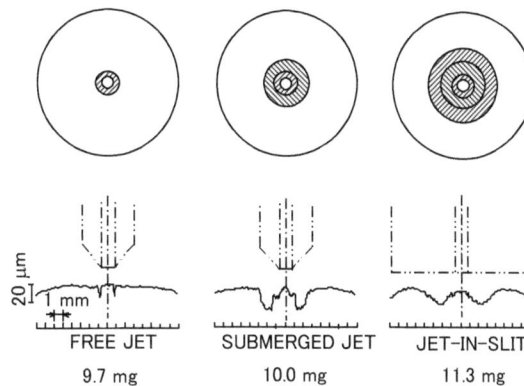

Fig. (4-3). Sketches of surface and cross section of specimen after 24 h test under flowing 3% sodium chloride solution [1].

Attack *a* occurred at same position on each specimen. Attack *b* was recognized on the specimens of submerged jet and jet-in-slit as well. In addition, attack *c* was recognized on jet-in-slit specimen. In the cross section, however, the distinction between *b* and *c* was not clear. According to the observation of scanning electron micrograph, the surface of the un-attacked area was covered with a thick oxide film. On the contrary the attacked surface was stripped of the surface film, exposing metal grains as well as the hill and valley structure on the grain surface. The test liquid used contained no solid particles, and it was confirmed that the fluid velocity was too low to develop cavitation. The localized corrosion of the test specimen under such conditions may be regarded as erosion-corrosion [1].

Reason of Generating Attack a: The textbooks of fluid dynamics indicate that shear stress τ_0 caused by fluid running on the surface of an object is defined by the following equation,

$$\tau_0 = \mu \, (du/dy)_{y=0} \tag{4-1}$$

where μ, viscosity of the fluid; $(du/dy)_{y=0}$, velocity gradient normal to surface (refer Fig. (**4-4**)).

As it was impossible to directly determine the velocity gradient in the close vicinity of the surface, an arrangement was used in order to measure the pressure difference that is proportional to the velocity gradient: holes of 0. 3 mm in diameter were provided on the surface at a 0. 5 mm interval and a hair was plastered in between (Fig. (**4-4**)). The principle on which τ_0 was determined by the pressure difference Δp across two holes is given by the following equation,

$$\tau_0 = \mu \ (du/dy)_{y=0} \propto u_1/y_1 \propto u_1 \tag{4-2}$$

$$\Delta p = p_1 - p_2 \propto u_1^2 \tag{4-3}$$

$$\tau_0 \propto \sqrt{\Delta p} \tag{4-4}$$

where y_1, a distance normal to the surface and smaller than the thickness of the boundary layer; u_1, fluid velocity at $y = y_1$.

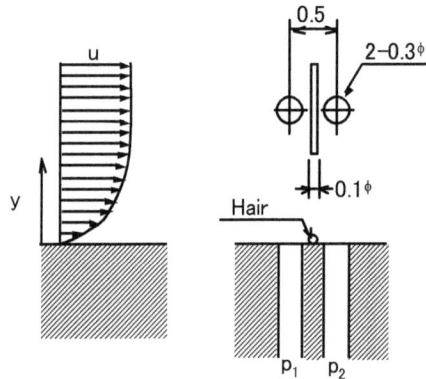

Fig. (4-4). Velocity gradient normal to surface wall and arrangement to measure the velocity gradient [1].

If the velocity gradient from wall surface to a distance y_1 is approximated to be constant, τ_0 is proportional to u_1/y_1 (Eq. (4-2)). On the other hand, the flow on the upstream side of the hair is prevented, which results in a pressure rise by the extent corresponding to dynamic pressure, that is, Δp (Eq. (4-3)). If we put y_1 equal to the diameter of the hair, we obtain Eq. (4-4). Although this method is inadequate for quantitative measurement because of many approximations involved, it indicates at least the situations on the surface where τ_0 is at maximum or zero.

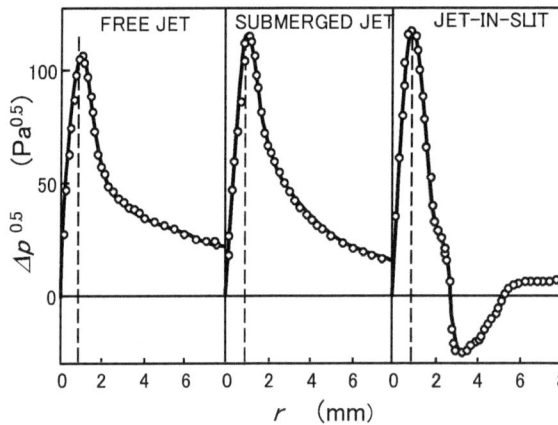

Fig. (4-5). Distributions of $(\Delta p)^{0.5}$ on each specimen surface [1].

The radial distributions of $(\Delta p)^{0.5}$ were measured by the arrangement (Fig. (4-4)). The results obtained for each nozzle are shown in Fig. (4-5). The dotted line in the diagram indicates the inside radius of nozzle. In each of the jet types, $(\Delta p)^{0.5}$ reaches its maximum near the perimeter of the jet and then falls as it moves toward the periphery of the specimen. The attack *a* in Fig. (4-3) is found on the cross section of each specimen where $(\Delta p)^{0.5}$ reaches maximum. Therefore, it may be confirmed that attack *a* was caused by shear stress.

Reason of Generating Attack b: In order to identify the cause of attack ***b***, we made a thin film (0. 1 mm thinness) of grease on a glass surface, placed it under the jet-in-slit nozzle, and watched the process of the film being broken away.

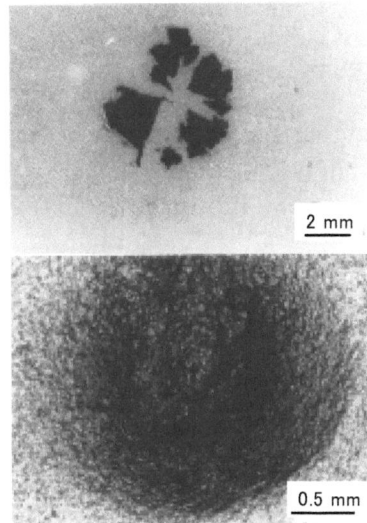

Fig. (4-6). Separation of grease film on a glass surface under jet-in-slit nozzle (upper) and behavior of grease film directly under nozzle (lower) [1].

Fig. (**4-6**), lower, is a photograph of the grease film directly under the jet, which shows that the grease film was continuously forced out by the shear stress toward the periphery. On the contrary, at the place where attack ***b*** or ***c*** occurs, the breakaway of the grease film progressed not continuously but suddenly, resulting in a small wedge shape at one time (Fig. (**4-6**)). This indicates that the force is an impulsive one that acts locally and momentarily. It is clear that the grease film and the oxide film are not the same, and the grease film cannot exactly simulate the oxide film. But it is insoluble in the solution. Thus, it proves that the grease film breakaway was caused not by dissolution but by mechanical force. It has also been made clear that the force is different from the shear stress.

The sudden separation of the grease film occurred on the submerged jet specimen as well, but no separation was recognized on the free jet specimen. So, we compare the flow conditions between the free jet and the jet-in-slit which are shown in Fig. (**4-2**): there is a clear difference in their flow velocity behavior between the jets just after they left the nozzle mouth; a steeper decrease in the radial flow velocity of the jet-in-slit than the free jet. This is of course due to the decrease in the fluid film depth of free jet shown in the figure, which brought about a more gentle increase in the cross section area of fluid flow, and accordingly a more gentle deceleration in the radial flow velocity of free jet than the jet-in-slit where film depth was kept constant at the distance of the slit. Important is that, as textbooks of fluid dynamics state, turbulence occurs most at the area where the fluid flow is decelerated. Thus, turbulence must have scarcely occurred in the free jet in accordance with the less deceleration in the flow.

In order to measure the intensity of turbulence in the fluid flow over the specimen, following arrangement was developed: a 0. 3 mm diameter pressure hole was provided at the root of a stainless steel tube, 0. 8 mm in outside diameter, and sealed at the top end (Fig. (**4-7 (1)**)). This pressure hole sensed the total pressure of the flow within 150 μm from the specimen surface, since only half of the hole was exposed to the fluid flow [2]. At the same time the static pressure was measured through another pressure hole on the specimen surface which was located near the total pressure tap (Fig. (**4-7(2)**)).

The total pressure, p_t, measured by the stainless tube sensor fluctuated considerably as shown in Fig. (**4-8**): The frequency was ca. 35 Hz. The amplitude of the fluctuation, Δp_t, must be an index representing the intensity of the turbulence, and was plotted against a converted distance from the center of nozzle since the measurement was carried out on an extension model of specimen of which diameter was 80 mm with a gap distance of 2 mm (Fig. (**4-9**)). It reached a maxim at the location of 2 mm downstream from the center of nozzle where attack **b** was originated. Therefore it may be related to turbulence of flow.

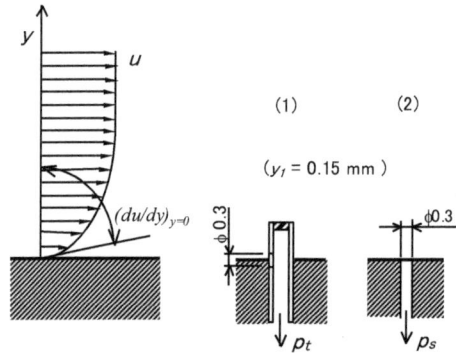

Fig. (4-7). Arrangements to measure (1) total pressure, and (2) static pressure [2].

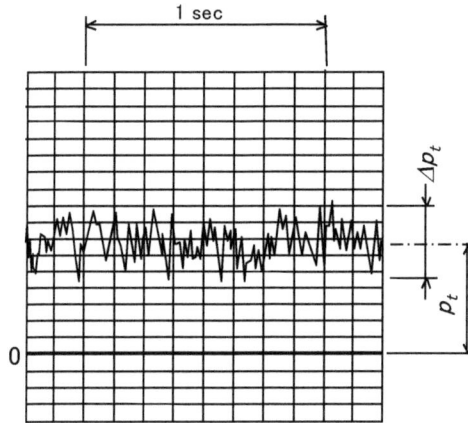

Fig. (4-8). Fluctuation of total pressure at the location where attack **b** occurs [2].

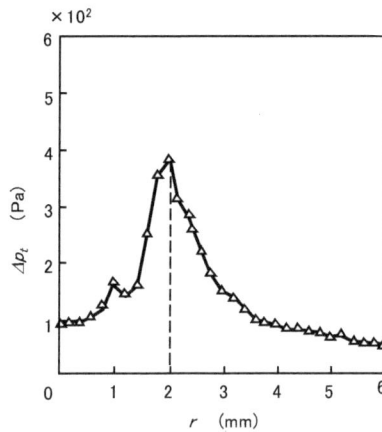

Fig. (4-9). Distribution of amplitude of total pressure fluctuation [2].

By the way, the dynamic pressure, p_d, was obtained by subtracting the static pressure from the total pressure. This is proportional to the square of the shear stress at the location, namely this has the same physical meaning with Δp in Eq. (4-4). The distribution of $p_d^{0.5}$ is shown in Fig. (**4-10**) as a function of the converted radial distance from the center of the nozzle. As shown in the figure, it reached a maximum at the location of about 1 mm downstream from the center of nozzle, which is in good consistence with Fig. (**4-5**). Thus, it was proven that two arrangements, those in Fig. (**4-4**) and Fig. (**4-7**) both function efficiently.

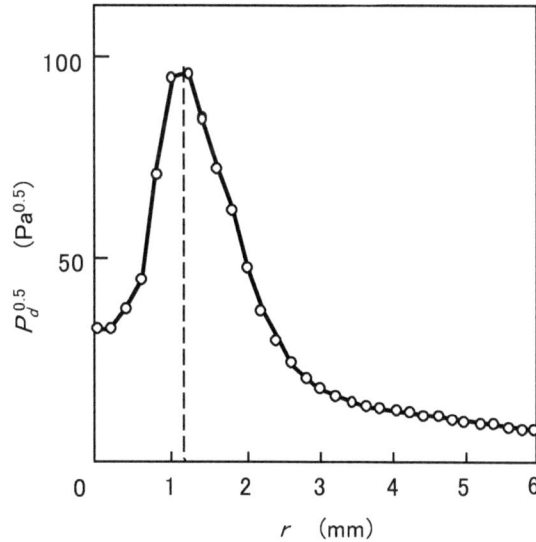

Fig. (4-10). Radial distribution of $p_d^{0.5}$ [2].

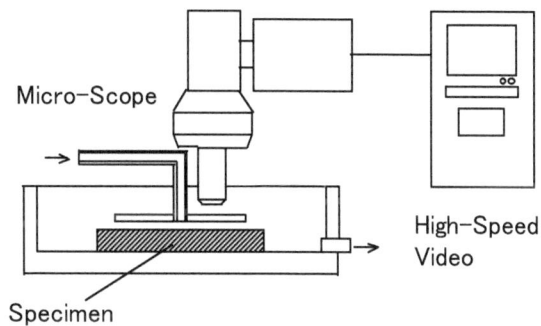

Fig. (4-11). Arrangements for observing oxide layer behavior.

You may be interested in how "turbulence" in fluid flow could separate the oxide layer from the base metal surface. An oxide layer was allowed to develop on a surface of brass specimen with the extension model size and it was put under a jet-in-slit. The behavior of the oxide layer was observed through a microscope and recorded in a high-speed video (Fig. (**4-11**)). It was revealed that the oxide layer consisted of coagulations (ca. 50 μm in diameter) of small particles several μm in diameter, which waved just like seaweeds in the ocean and were then swept away. The frequency of the waving was about 35 to 40 Hz, which coincided satisfactorily with that of p_t in Fig. (**4-8**).

The flow geometry of submerged jet resembles that of impingement attack in Fig. (**1-10**) so well that it must possibly reproduce this type of erosion-corrosion on the specimen. It is also quite feasible that the attack ***a*** is caused by the shear force and the attack ***b*** by the turbulence force, the latter being by far more powerful than the former.

In order to compare those forces in contribution to cause inlet tube corrosion, a tube of the 6/4 yellow brass mentioned above with 8 mm in inside diameter was prepared and a 3 % salt solution (40 °C) was allowed to flow through it with a flow velocity of ca. 5 m sec^{-1}. After 24 h corrosion damage of ca. 6 mm in width arose at ca. 7 mm downstream from the entrance of the tube. Total pressure and static pressure were measured using the arrangements shown in Fig. (4-7) at several locations along the tube, and the distribution of dynamic pressure, that is p_d proportional to the square of the shear stress and Δp_t proportional to the turbulent force were determined (Fig. (4-12)): the former decreased monotonously from the entrance of the tube but the latter, the representative of turbulent force reached a maximum in the region where the localized corrosion arose. In conclusion, inlet tube corrosion shown in Fig. (1-10) as a form of erosion-corrosion may be attributed not to shear force but rather to turbulence force.

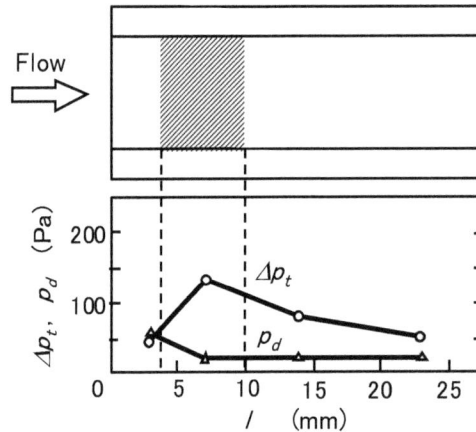

Fig. (4-12). Distribution of p_t and Δp_t along the axis of tube: bore, 8 mm; flow velocity, 5 m sec^{-1} [2].

1.3. Jet-in-Slit with Reverse Flow

Another type of jet-in-slit was developed: as illustrated with the solid arrows in Fig. (4-13), so far the test solution was allowed to impinge at a right angel to the specimen (the lower disc) and then flow in a radial direction through the slit; a new idea was to reverse the flow direction of the test liquid as indicated with the open arrows; the test solution was aspirated from the periphery up to the nozzle.

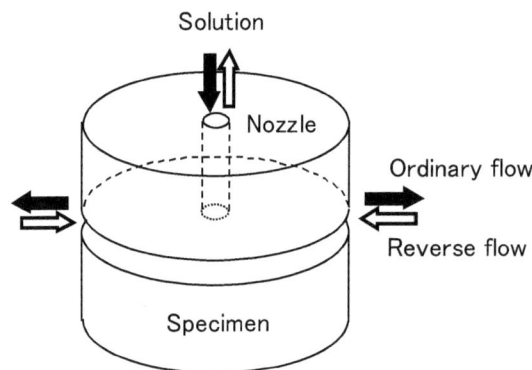

Fig. (4-13). Jet-in-slit with ordinary flow or reverse flow.

The idea came from textbooks of hydrodynamics as mentioned above: turbulence in flow is enhanced by the deceleration of flow velocity but suppressed by the acceleration. In the jet-in-slit with reverse flow no deceleration occurs and consequently there must be no turbulence. Thus, it was expected that the effect of the velocity gradient on the corrosion rate must be clearly examined without the influence of turbulence. Flow patterns in the slit were visualized by pinning threads on the specimen surface, one at the center of ordinary flow specimen, another at the periphery of reverse flow specimen (Fig. (4-14)). As expected, the thread in the ordinary flow bent greatly and vibrated intensely at the location where the intensity of turbulence reached its peak and the attack *b* occurred. In contrast, the thread pinned in the reverse flow was straight and did not vibrate at all.

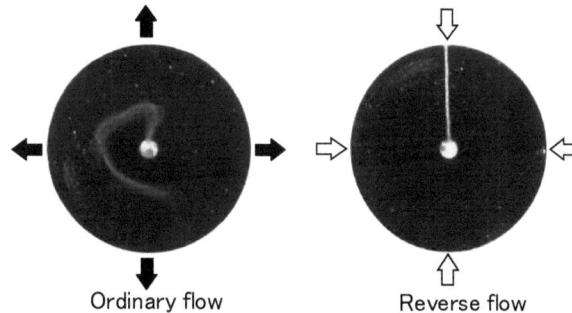

Fig. (4-14). Visualization of flow pattern in slit: flow rate, 0. 8 L/min; shutter speed, 1/15 sec.

Fig. (**4-15**) shows cross sections of pure copper specimens after a 1 hr reverse flow test in a 1% copper chloride (CuCl$_2$) solution (as described later in Chapter 5 this solution gives no influence on the corrosion mechanism). At a lower flow rate (0. 20 L min^{-1}) a small swelling appeared to have arisen in the center of the specimen (upper cross-section). This, of course, was not true. The fact was that the higher dissolution rate at the periphery of specimen made the central portion stand out where the dissolution rate is lower than that of the periphery in spite of higher flow velocity. At a higher flow rate (0. 80 L min^{-1}), dramatic inversion occurred, that is, a remarkably deep damage appeared at the center of the specimen (lower cross section).

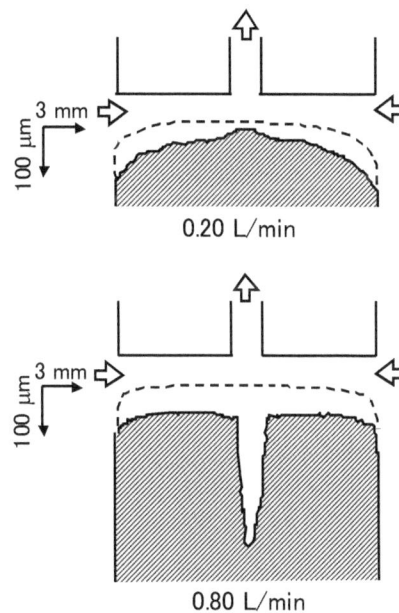

Fig. (4-15). Profiles of corrosion damage originated in jet-in-slit with reverse flow: specimen, pure copper; test liquid, 1 % CuCl$_2$ solution [3].

In order to find out a good reason for the exchange from the swelling to the deep hole, the liquid flow was visualized by putting aluminum powder in the liquid, and flow patterns at the entrance to the nozzle were observed. As a result, it was found that at the higher flow rate a fixed vortex covered the specimen surface just under the nozzle mouth (Fig. (**4-16**)). It is well known that a fixed vortex involves the stagnant watershed where the liquid is seldom renewed, so that the dissolved oxygen content may be lower than that in the main flow. The oxygen supply to the surface beneath the vortex must be poor, and consequently a coarse oxide layer with a poor barrier against the migration of base metal ions must have formed there.

Lower flow rate

Higher flow rate

Fig. (4-16). Liquid flow visualization at nozzle mouth in jet-in-slit with reverse flow [3].

Combining the results of test at lower and higher flow rate, we may conclude that a sort of erosion-corrosion which is similar to the differential flow-velocity corrosion of cast iron described in Chapter 1 was reproduced on copper alloy in a jet-in-slit test with reverse flow. In chorus with this we cannot but recognize the important role of the fixed vortex played in the reproduction of this sort of erosion-corrosion since it make us notice the possibility that the horseshoe corrosion shown in Fig. (1-10) originated from a fixed vortex as well: the image of banana vortex in Fig. (4-17) is believed to be the origin of the turbulence in fluid flow over an obstacle: a more accurate structure of the vortex can be imaged from the characteristic pattern of horseshoe corrosion as shown in Fig. (4-18).

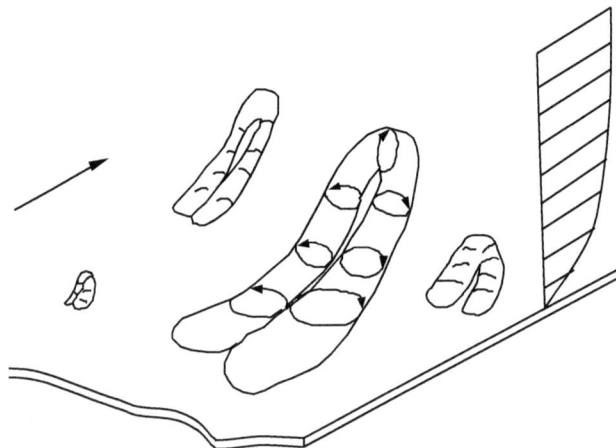

Fig. (4-17). Image of banana vortex proposed by Y. Fukunishi and H. Sato as the origin of turbulence in fluid flow [3].

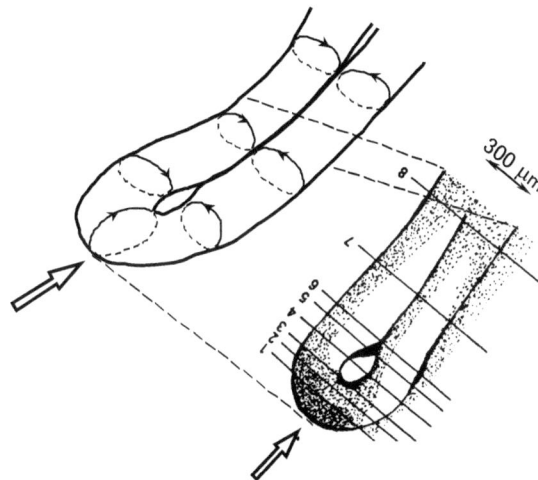

Fig. (4-18). **Vortex** imaged from the shape of banana vortex and characteristic pattern of horseshoe corrosion, and presumed to be the origin of horseshoe corrosion.

2. THEORIES OF ELECTROCHEMICAL CORROSION IN FLOWING FLUID

2.1. Introduction

In the previous section it has been revealed that the jet-in-slit test with ordinary and reverse flow seemingly reproduces some characteristic damage pattern of erosion-corrosion on the specimen surface. It also produces unique flow patterns as well, and each pattern of attack or damage was rather reasonably related to each flow pattern. We could understand some of the relationships between them, for example, a high shear force at high flow velocity area would possibly cause separation of oxide layer exposing the underlying metal surface to the environment which would result in localized corrosion damage. However, the reason why a deep pit was originated under a fixed vortex, or the reason why larger thinning occurred at the area of lower flow velocity than the area of higher flow velocity was not given yet. Is there any unified theory which reasonably correlates every damage pattern to each of flow pattern? Answer to this question ought to be found from the theory of pure electrochemical corrosion which was appointed to be the true nature of erosion-corrosion in the previous chapter.

2.2. Models of Corrosion Process and Mass Balance

Fig. **4-19** shows the definition of uniform corrosion and localized corrosion.

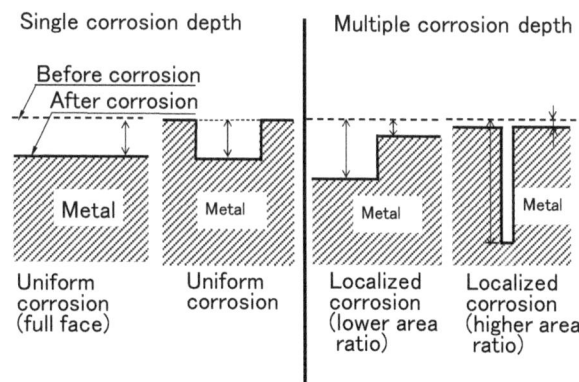

Fig. (4-19). Illustration for the definition of uniform corrosion and localized corrosion.

A uniform corrosion is defined as a single depth of corrosion over all the surface of metal. For a localized corrosion, on the other hand, multiple (at least two) corrosion depths are to be on the surface. Therefore, the micro-cell corrosion model with a single anodic dissolution rate is for uniform corrosion and the macro-cell corrosion model with multiple (at least two) anodic dissolution rates for localized corrosion.

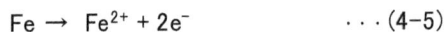

$$Fe \rightarrow Fe^{2+} + 2e^- \qquad \cdots (4\text{-}5)$$

(Oxidation reaction, anode reaction)

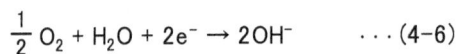

$$\frac{1}{2} O_2 + H_2O + 2e^- \rightarrow 2OH^- \qquad \cdots (4\text{-}6)$$

(Reduction reaction, cathode reaction)

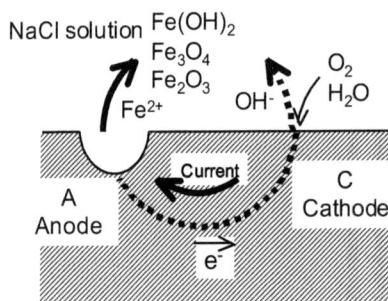

Fig. (4-20). Micro-cell model [4].

In Fig. (**4-20**) the micro-cell corrosion model is applied, as usual in textbooks, to a piece of iron metal in an aqueous environment which contains dissolved oxygen [4]. At the anode, the metal is oxidized into Fe^{2+} ions losing electrons (Eq. (4-5)). Fe^{2+} ions dissolve in environment resulting in a loss of metal at that location. Electrons migrate through the metal to reach cathode where they are consumed *via* the reduction of oxygen into hydroxide ions (Eq. (4-6)). Here, no mass loss of metal occurs. Fe^{2+} ions and hydroxide ions react to give iron hydroxide, $Fe(OH)_2$, which is oxidized to magnetite, Fe_3O_4, and then further to hematite, Fe_2O_3. The accumulation of the latter two corrosion products over the metal surface composes so-called protective corrosion product films, the characteristics of which may control the migration rate of metal ions from the metal surface to the bulk of the environment. The details of these films are, however, not important for the present discussion.

It is important in the micro-cell model that electrons migrate from the anode to the cathode through the metal. Electrons are negatively charged, so that an electric current flows in the direction opposite to that in which electrons migrate. The electrons remain within the metal, as does the current. Fe^{2+} ions, contrary to this, leave the metal and enter the environment carrying a positive charge which is equivalent in amount to that of the electrons produced in the reaction (Eq. (4-5)). This implies that an electric current equivalent to the in metal flowing current flows from the anode to the environment. This is referred to as anodic current. On the other hand, hydroxide ions leave the cathode, entering the environment carrying a negative charge, resulting in an electric current that flows from the environment to the cathode. This is referred to as cathodic current. Thus, the electric current caused by the migration of electrons within the metal is connected with those caused by the migration of ions to complete an electric circuit which is apparently similar to that of a typical battery. The structure, which consists of anode, cathode and environment, is, accordingly, called a corrosion cell. In electrochemistry, a piece of metal placed in an aqueous environment is called an electrode or sometimes a composite electrode when anode and cathode are simultaneously present on a metal surface at the same time. In this connection, the anode or cathode alone is called a single electrode or half cell.

The micro-cell corrosion process shown in Fig. (**4-20**) as well as every chemical reaction process must obey the law of conservation of mass. We can see the law holds in the corrosion process since the reaction rate of the cathode is balanced with that of the anode, each being equal to the current which flows between them. The chemical reaction rate in the amount of moles per unit time is converted into equivalents per unit time by multiplying the amount of moles by the numeral of charges of the ion concerned. The Faraday constant, 96500 coulomb/equivalent, is used to convert it further into coulombs per unit time, that is, amperes [A]. In this connection, the rate of an anode reaction is often called the anodic current and the cathode reaction rate the cathodic current. In the case of multiple reduction reaction, that is, when the reduction of hydrogen ion parallels oxygen reduction, the total amount of reduction must be equal to the total oxidation at the anode.

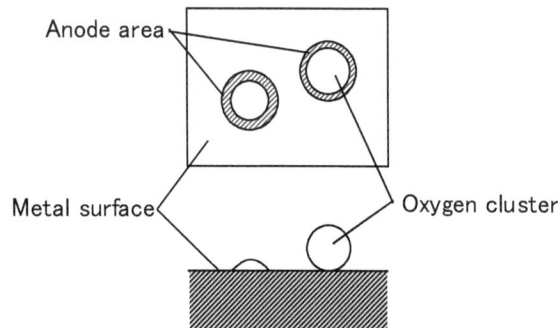

Fig. (4-21). Image of oxygen cluster [4].

One of the shortcomings of the micro-cell corrosion model is that it looks to be different from the image of the uniform corrosion: corrosion damage appears localized at the anode. The usual explanation that the anode and the cathode exchange their positions from time to time to result in uniform damage is regretfully far from persuasive. The author's illustration for this type of corrosion is given in Fig. (4-21): oxygen dissolved in water may not be uniformly dispersed but possibly coagulates to form clusters with a certain dimension. When an oxygen cluster contacts the metal surface, the reduction of oxygen begins by obtaining electrons from the surface. At the same time, the oxidization of the metal take place on the metal surface which surrounds the area in contact with the oxygen cluster. The micro-cell thus constituted vanishes as soon as all the oxygen in the cluster is reduced. Here again, oxidation and reduction are balanced. Such micro-cells with a short life time occur successively and uniformly over the metal surface resulting in a uniform loss of material. This time-depending process of uniform corrosion cannot be illustrated in a single drawing, so it was symbolized with a metal surface which is composed of numerous micro-cells, as shown in Fig. (4-22), which is referred to as multi-cell for convenience.

Fig. (4-22). Multi-cell model [4].

It is a matter of course that the multi-cell model cannot illustrate the mechanism of localized corrosion, nor can the micro-cell model because in the real corrosion of this type (localized corrosion) a more or less of mass loss occurs in the cathode area, which beyond the scope of the micro-cell model. Fig. (4-23) shows the macro-cell model which was introduced for illustrating localized corrosion. It consists of two micro-cells (actually multi-cells): the left side cell for macro-anode and the right side for macro-cathode. In each multi-cell, the anodic current and cathodic current is not equal to each other, and oxidation and reduction are not in balance. Nevertheless, macro-cell current which is flowing from the macro-cathode to the macro-anode maintains the mass balance in the macro-cell, that is, the total anodic current is equal to the total cathodic current.

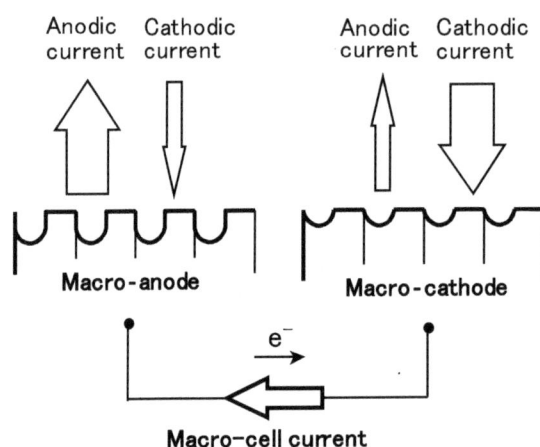

Fig. (4-23). Macro-cell model [4].

The common drawback in the preceding three corrosion models is that the surface area of electrode is not taken into consideration. The most practical unit of corrosion rate is the rate of wall thinning in [mm y^{-1}] or [mm^3 mm^{-2}y^{-1}]. Metallic materials for engineering use are classified according to their corrosion rates in this unit: those with the rates < 0. 1 mm y^{-1} are judged usable, and those >1 mm y^{-1} not usable. To obtain the wall thinning rate in mm y^{-1}, the mass loss rate of the metal due to corrosion is divided by the density of the metal and the surface area where the mass loss originates. This corresponds to the current density, [A m^{-2}]. On the other hand, the oxidation *vs.* reduction balance must be discussed in terms of electric quantity, coulombs, or current, coulomb sec^{-1}, that is amperes [A]. In order to avoid confusion, current density is adopted as the measure of corrosion rate for the present under the assumption that the macro cathode area is equal to the macro anode area. The surface area of electrodes will be taken into consideration when it is necessary.

2.3. Driving Force for Corrosion Process

As mentioned above, the mass balance in a uniform corrosion process is presented by the micro-cell model and that of a localized corrosion process by the macro-cell model. The macro-cell model, however, is not sufficient to present a detailed mechanism of any type of localized corrosion: it cannot distinguish between the differential flow- velocity corrosion and the impingement attack. To do this, a potential *vs.* current density diagram must be used. The potential has much to do with the driving force for promoting corrosion processes and the current density is exactly the rate of corrosion as described above. This diagram therefore achieves a complete description of the mechanism of corrosion process. Here, potential means a type of energy level of electrons or the concentration of electrons in a piece of metal, in the same way that the chemical potential of the ions in the environment means the concentration of ion, Fe^{2+} for example. The reason why the potential functions as the driving force in a corrosion process is that the electrochemical reaction equation, Eq. (4-5) for example, is reversible reaction.

$$Fe \leftrightarrow Fe^{2+} + 2e^- \hspace{6cm} (4\text{-}5)'$$

In general, a reversible reaction proceeds in a direction from the higher potential side to the lower potential side. When the total potential in each side of the equation are in balance, the rate of the rightward reaction equals with that of the leftward so that no apparent reaction proceeds. This state of reaction rates is called equilibrium. The potential of a metal depends only on the temperature and that of ion on the temperature and its concentration in the environment as well. Referring to Eq. (4-5)', under the conditions where the concentration of Fe^{2+} as well as the temperature of the environment is kept constant, only the potential of the electron is variable, so that it controls the potential balance of the equation. Since electrons are negatively charged, a higher level of electron energy corresponds to a lower value of the potential. Concerning the direction of the reaction, it proceeds left-ward (reduction proceeds) when the potential is lower than that at the equilibrium, since the total energy of the right hand side of the equation is higher than that at equilibrium, and *vice versa* (oxidation proceeds). Concerning the rate of the reaction, the intensity of the reaction-driving force totally depends on how far the potential deviates from the equilibrium potential, because the reaction rate in any direction is zero at equilibrium. In the electrochemistry, the deviation in potential from the equilibrium potential is called polarization. Namely, polarization is the major driving force of a reversible electrochemical reaction, and a potential *vs.* current density relation is usually referred to as polarization curve.

2.4. Resistance to Electrochemical Reaction

Fig. (**4-24**) shows an anodic polarization curve in a potential *vs.* current density diagram. The deviation in potential from the equilibrium potential is the anodic polarization, that is, the anode reaction driving force. The current density on the abscissas axis in a logarithmic scale is the measure of reaction to proceed. An anodic polarization curve provides useful information on the corrosion behavior of a metal just as a stress *vs.* strain curve does in presenting the strength of material. It consists of several parts which may be approximated by straight lines. The slopes of these lines have the dimension of volts [V], meaning the potential difference necessary for increasing the reaction rate by 10 times. In the electrochemistry this is referred to as polarization resistance as it represent the level of resistance to accelerating the corresponding electrode reaction.

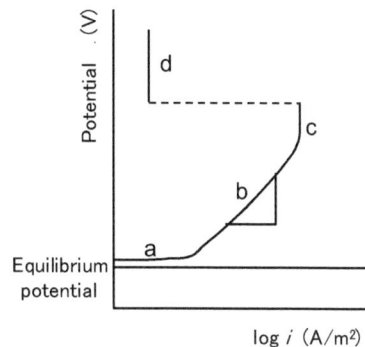

Fig. (4-24). Anodic polarization curve (schematic) [4].

Part **a** in the anodic polarization curve is under the influence of exchange current which is flowing at the equilibrium in directions, oxidation and reduction, in the same time with the same rate. The larger the current, the longer this portion of the curve is. Part **b** is called activation polarization which is derived from the activation energy required to promote a chemical reaction, the transfer of electric charge in this case. Part **c** is called concentration polarization. This sort of polarization is derived from the storage of corrosion products on the anode surface: with the increase in the concentration of Fe^{2+} ion of Eq. (4-5), the chemical potential of the ion increases, which suppresses the reaction by raising the chemical potential of the right-hand side of the equation. This type of polarization also exists in part **b** which is not conspicuous, since the storage of ions is rather small as the current density is low. It would, however, be conspicuous in part **c** since the corrosion products film composed of magnetite, Fe_3O_4, is developed on the metal surface, which prevents Fe^{2+} ions migrating to the bulk of environmental fluid.

In part **d**, the passivity of iron is achieved where the thick corrosion products film is replaced by a much thinner but strong film: when the potential reaches at a characteristic potential called Flade potential, the current, that is, the reaction rate suddenly drops to a level several orders of magnitude lower. Flade potential is a function of the temperature and pH of the environment as well as of the type of metal being considered. Stainless steels have lower Flade potential values, and therefore are passive at room temperature presenting excellent corrosion resistance.

A cathodic polarization curve is shown in Fig. (**4-25**). The deviation of potential from equilibrium to the negative direction is the cathode reaction promoting force. Although the direction of the current is also negative, it is placed on the abscissas axis in the order of increase as it proceeds rightward. This is for convenience in comparing it with the anodic current. As this description way is the proposal of Evans, the diagram of this mode is generally called Evans diagram.

Part **a** as well as part **b** of the cathodic polarization curve is derived from a similar origin as that of the anodic curve. Similarly, part **c** is also referred to as concentration polarization. It is derived, however, not from the increase in concentration of ions, but from the exhaust of oxygen near the cathode area. As described later, oxygen is transported from the balk to the metal surface through diffusion. So that, once the concentration of oxygen became zero, the rate of oxygen reduction or the cathodic current becomes equal to the mass transfer rate of oxygen which depends on the fluid flow conditions over the surface but is independent of potential. Regardless of a further increase in the reaction driving force, the current remains at the level corresponding to the oxygen diffusion rate. This is referred to as oxygen diffusion limiting current density.

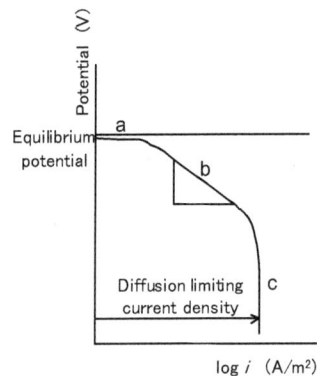

Fig. (4-25). Cathodic polarization curve (schematic) [4].

2.5. Illustration of Uniform Corrosion Mechanism with Evans Diagram

For a metal corroding in the environment, next three conditions must be satisfied.

1. The total amount of reduction is equal to the total amount of oxidation.

2. The potential is the same over all the metal.

3. Concerning single electrodes, oxidation proceeds at a higher potential than that at equilibrium, so does reduction at a potential lower than that.

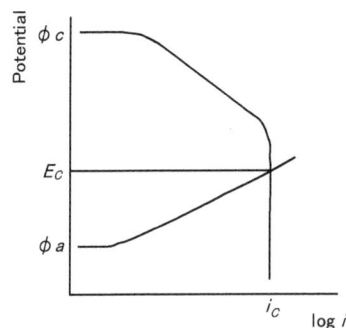

Fig. (4-26). Evans diagram for micro-cell (actually multi-cell) model [4].

The preceding conditions are all satisfied at the point where the anodic and cathodic polarization curves cross each other. The coordinate position of the cross point gives the corrosion potential, Ec, and the corrosion current density, i_c as shown in Fig. (**4-26**). In this case, the corrosion current density is equal to the oxygen diffusion limiting current density, which derived from the situation in which the resistance to cathodic polarization is larger than that to anodic polarization. Since these single electrodes are arranged in series in the corrosion cell circuit, the electrode with a larger resistance controls the circuit current. This situation shown in Fig. (**4-26**) is called cathodic control. Under a complete cathodic control, metal A and B shows the equal corrosion rate, which is the oxygen diffusion limiting current, regardless the difference in the characteristics of the metals: metal A is a noble metal which has a higher equilibrium potential as well as a larger anodic polarization resistance; metal B is a base metal with a lower equilibrium potential and less resistance to anodic polarization (Fig. (**4-27**)).

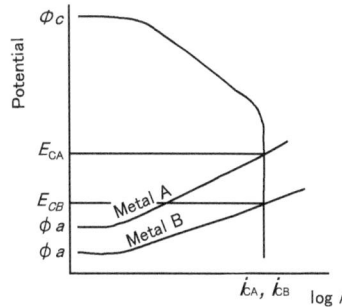

Fig. (4-27). Corrosion rates of different sort of metals under cathodic control [4].

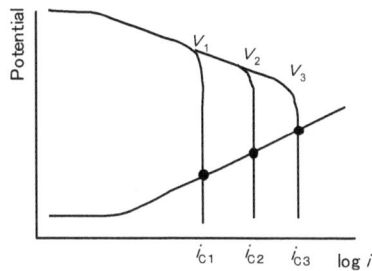

Fig. (4-28). Influence of flow velocity on corrosion rate [4].

Under the cathodic control, the corrosion rate of metal in flowing solution depends on the flow velocity as shown in Fig. (**4-28**). The reason for this is illustrated in Fig. (**4-29**) where the concentration distribution of oxygen dissolved in the environmental fluid near the metal surface is given. In the bulk fluid, the oxygen distribution is uniform as the fluid is mixed well by convection. Close to the metal surface, the fluid velocity is low and fluid mixing is minimal. This is called boundary layer.

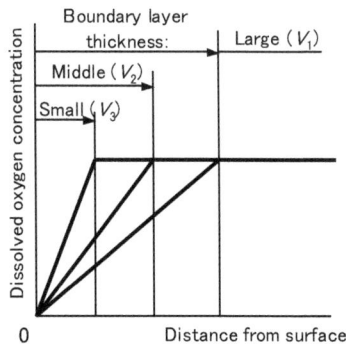

Fig. (4-29). Distribution of dissolved oxygen concentration in the vicinity of a metal surface [4].

On the metal surface, the oxygen concentration is zero because it is reduced to hydroxide ion as soon as it reaches the surface. As a consequence, the oxygen concentration distribution in boundary layer becomes a straight line which connects zero at the metal surface to that at the surface of boundary layer. The straight line implies that oxygen is transferred through diffusion in the layer and the rate of transfer, N, is given by the following equation,

$$N = D \, (dC/dX) \tag{4-7}$$

where D is the diffusion coefficient, C is oxygen concentration and X is the distance from the metal surface. At constant D, N is proportional to (dC/dX) which corresponds to the slope of the straight lines in Fig. (4-29). As the flow velocity increases in the order $V_1 < V_2 < V_3$, the thickness of the boundary layer decreases in the order $V_1 > V_2 > V_3$, which makes the slope of the straight line steeper and the oxygen diffusion limiting current increase in the same order $V_1 < V_2 < V_3$. Thus, the corrosion rate is increased with increasing flow velocity.

Fig. (4-30) shows the corrosion potential and corrosion rate in a situation where hydrogen ion reduction parallels oxygen reduction. As mentioned above, the corrosion potential is given by the condition where the total anodic current is equal to the total cathodic current. Under the assumption of a common electrode surface area, the current may be replaced by current density. So that $i_l + i_h = i_c$ is established in Fig. (4-30). Though it may be not seen like that, the relation is certainly established: remember that the abscissa axis is expressed on a logarithm scale.

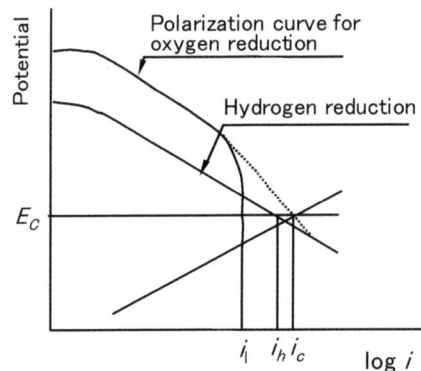

Fig. (4-30). Corrosion potential and corrosion current in the case where hydrogen ion reduction parallels oxygen reduction: dotted line, the sum of reduction reaction rates [4].

2.6. Illustration of Localized Corrosion Mechanism with Evans Diagram

The Evans diagram (potential *vs.* current density diagram) in Fig. (4-31) illustrates the mechanism of so-called galvanic corrosion based on the macro-cell model applied to two different metals. At the same time, it illustrates the mechanism of cathodic protection with sacrificial anode. The corrosion potentials of a noble metal and a base metal with a common surface area are different from each other as shown by the solid dots in the diagram, but they show an identical corrosion rate i_l, that is, the oxygen diffusion limiting current density.

When these metals are connected through a conducting wire, the corrosion rate of the noble metal decreases to i_P, and that of the base metal increases to i_S. This is because the potentials of the metals move to the identical level as shown by the open dots where the total cathdic current, $(i_l + i_l)$, is equal to the total anodic current, $(i_P + i_S)$. This illustration using Evans diagram may be applied to the case of copper in contact with iron: by the contact, the potential of copper decreases with a decrease in corrosion rate to some extent but is not terminated. In contrast, the potential for iron increases and the corrosion rate increases to a level higher than when it is independently corroded. Iron plays the role of a sacrificial anode and, at the same time, the role of the macro-anode of a macro-cell resulting from the contact of the metals.

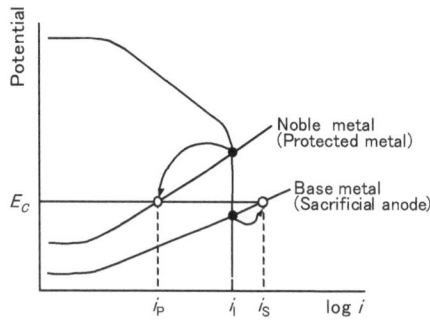

Fig. (4-31). Formation mechanism of so-called galvanic corrosion [4].

Fig. (**4-32**) shows an example in which the polarization curves on the diagram fail to illustrate the mechanism of differential aeration-cell corrosion or differential oxygen concentration-cell corrosion. On a metal surface, two zones are assumed to be present: one of which is exposed to a fluid flow with a lower velocity and with a poor oxygen supply: the other to a higher velocity with affluent oxygen supply (*L* and *H* in Fig. (**4-32**)). As these zones are really short-circuited with each other through the bulk of the metal, the origination of a macro-cell, consisting of a macro-anode with a higher dissolution rate and a macro-cathode with a lower dissolution rate would be expected. Contrary to expectation, the corrosion rate had an equal value i_c resulting not in localized but in uniform corrosion.

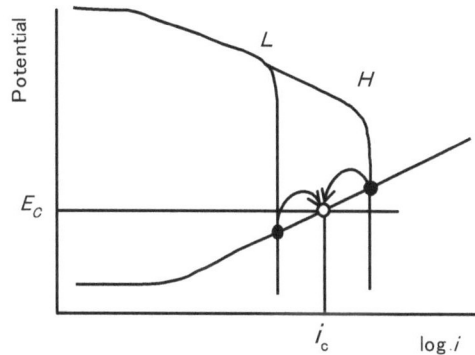

Fig. (4-32). Example failed in illustrating the mechanism of differential aeration-cell corrosion [4].

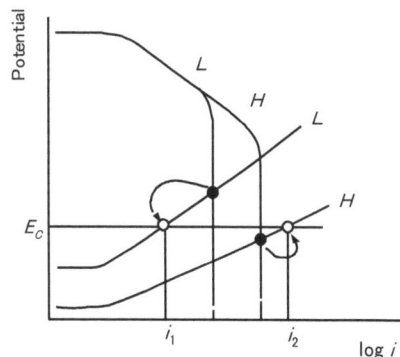

Fig. (4-33). Evans diagram illustrating localized corrosion originated in jet-in-slit with ordinary flow [4].

In Fig. (**4-33**) other assumptions were added to that of Fig. (**4-32**): it was assumed that a fluid flow of higher velocity or a higher level of turbulence at zone *H* may remove the protective oxide layer from the surface or must at least clear the positive ions from the electrode surface to substantially decrease the anodic polarization resistance. Concerning cathodic polarization, the assumption remains the same as that of Fig. (**4-32**): the oxygen diffusion limiting current density at zone H is higher because of the higher oxygen supply rate there.

The combination of zones H and L composes a macro-cell where zone H becomes the macro-anode, and the metal is oxidized at that location at the rate i_2, which is higher than when the zone corrodes independently. In contrast, zone L becomes a macro-cathode, and metal is oxidized at the rate i_1 which is lower than when it corrodes independently. This is an illustration of the localized corrosion originated in the jet-in-slit test with ordinary flow, and accordingly the impingement attack and inlet tube corrosion in Fig. (**1-10**).

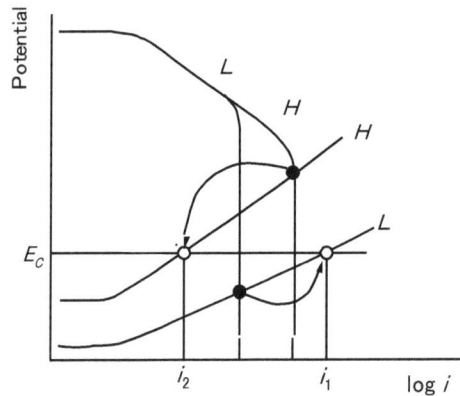

Fig. (4-34). Evans diagram illustrating localized corrosion originated in jet-in-slit with reverse flow [4].

In Fig. (**4-34**) an alternative assumption was added to that of in Fig. (**4-32**): a higher anodic polarization resistance for zone H. This can occur when a dense and stable corrosion product film of excellent protective qualities covers the surface due to a sufficient supply of oxygen as well as anions which consist of the film cooperating with oxygen. Alternatively, a lower anodic polarization resistance for zone L may be when it is covered with a fixed vortex: poor supply of oxygen may result in a thin and coarse film with poor protection, accordingly a lower anodic polarization resistance for zone L. In both cases the combination of zones composes a macro-cell where zone H is the macro-cathode and zone L the macro-anode, where the metal is dissolved at a rate higher than that of the oxygen diffusion limiting current at the macro-cathode. This is the illustration of the localized corrosion originated in the jet-in-slit test with reverse flow, that is, differential flow-velocity corrosion, horseshoe corrosion as well as deposit attack.

2.7. Advantage of Evans diagram in Description of Corrosion Mechanism

In comparing Figs (**4-33**) and (**4-34**), it can be seen that the relative position of the cathodic polarization curves, L and H, in the diagrams are the same, but that of the anodic curves are reversed. Thus, the relative position of the anodic polarization curves discriminates the localized corrosion types which are originated in jet-in-slit with ordinary flow and that with reverse flow. A feature common to these types of localized corrosion is that the current density at the macro-anode is much higher than the highest oxygen diffusion limiting current. The oxygen diffusion limiting current is, as the name implies, the maximum rate that oxygen transfer through boundary layer can reach but cannot exceed. It is also the maximum among the corrosion rates when the zones in Figs. (**4-33**) and (**4-34**) corrode independently under cathodic control. Once a macro-cell is formed, however, the maximum is simply exceeded by the corrosion rate at the macro-anode. The extent of the increase in corrosion rate is larger than it appears since the current density on the abscissas is given on a logarithmic scale.

It has been described in Chapter 1 that the rate of carbon steel pipe wall thinning caused by erosion-corrosion so far reached 9 mm y^{-1} at maximum. This attains to 90 times the corrosion rate that is allowed in industry. Such a high wall thinning rate in the field cannot be explained without assuming not only the formation of macro-cell but also a far larger surface area of macro-cathode as compared to that of macro-anode. As pointed out in a preceding section, the mass balance for a corrosion cell must be achieved not in current density but in current. The decrease and increase shown by the narrow arrows in Figs (**4-33**) and (**4-34**) must be balanced accordingly in the current. These currents are divided by the electrode surface area to obtain the current densities which correspond to wall thinning rates. Assuming that the ratio of the macro-cathode area to the macro-anode area is 10 to 1, the current density at the macro-anode in Figs (**4-33**) and (**4-34**) must be multiplied by 10 to obtain the actual wall thinning rate.

In conclusion, based on a **micro**-cell of corrosion the anodic dissolution rate or the wall thinning rate may be increased with the increase in flow velocity since the oxygen transfer rate in boundary layer increases with flow velocity, but it can never exceed the oxygen diffusion limiting current. Nevertheless, once a **macro**-cell of corrosion is formed the wall thinning rate can increase beyond the limiting current in consequence of the macro-cell current. Even with the formation of a macro-cell, however, such an unusual high rate of wall thinning, 9 mm y^{-1}, would not be comprehended without considering the surface ratio of macro-cathode to macro-anode. This to be called surface area ratio effect is the important factor which ranges with the macro-cell current effect.

2.8. Demonstration Experiments

Murakami who was a graduate student, given excellent talent and technique conducted experiments to prove that the characteristic corrosion patterns produced in jet-in-slit with reverse flow as shown in Fig. (**4-15**) was brought about through the formation of macro-cell of corrosion regardless the flow rate [3].

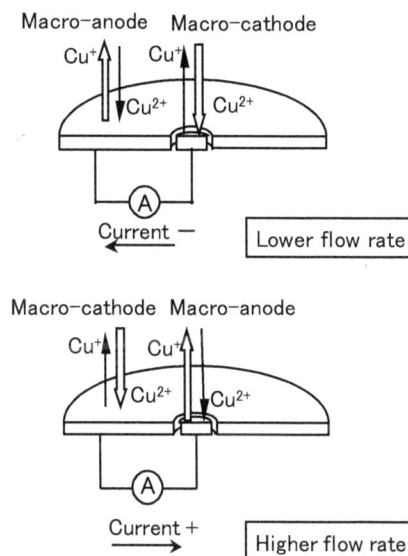

Fig. (4-35). Macro-cell corrosion models in jet-in-slit with reverse flow [3].

First, he visualized a macro-cell model following the Evans diagram in Fig. (**4-34**) as shown in Fig. (**4-35**). At a lower flow rate, the flow velocity of the central part is higher than that of the periphery, which results in higher diffusion rate of Cu^{2+} and Cl$^-$ from the bulk flow to the metal surface developing there a protective oxide layer preventing the migration and consequently suppressing the dissolution of metal ions, so that the central part becomes a macro-cathode. Accordingly, the surface of the peripheral part becomes a macro-anode which emits an electron more easily than the central part. The surplus electrons move towards the central part, and equivalent electric current flows toward the periphery. This is the macro-cell current (the narrow allows in Fig. (**4-34**)).

In contrast, at a higher flow rate, a vortex is fixed at the central part, beneath which the flow velocity is very low. At the peripheral part, the flow velocity is higher than that beneath the vortex, and a larger amount of Cu^{2+} and Cl^- ions is supplied from the bulk of flow to the surface. A good protective film develops there suppressing the dissolution of metal ions so that this part becomes a macro-cathode, alternatively the central part a macro-anode. A macro-cell current flows into the macro-anode as shown in Fig. (**4-34**) accelerating the corrosion rate in the central part.

In order to verify the above model the macro-cell current was measured at the lower and a higher flow rate in the jet-in-slit with reverse flow. The central part to the specimen was insulated from the peripheral part, and a zero shunt ammeter was connected between the two parts. The variations in the macro-cell current with time are shown in Fig. (**4-36**). The directions of the positive/negative of macro-cell current were defined in Fig. (**4-35**). At the lower flow rate, the ammeter indicated a negative current as expected. The amount of the current, however, was as small as about 1 μA. On the other hand, at the higher flow rate a positive current with substantial amount, 200 μA, was detected. Thus, the macro-cell currents flowed in those directions as predicted by the model.

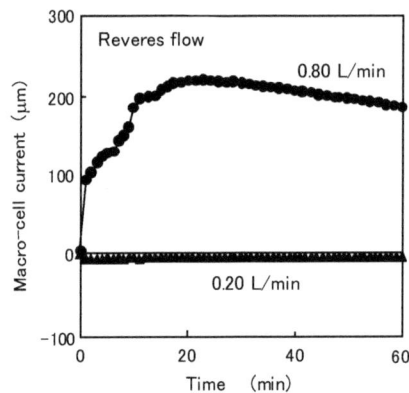

Fig. (4-36). Macro-cell current measured through a zero shunt ammeter during jet-in-slit test with reverse flow [3].

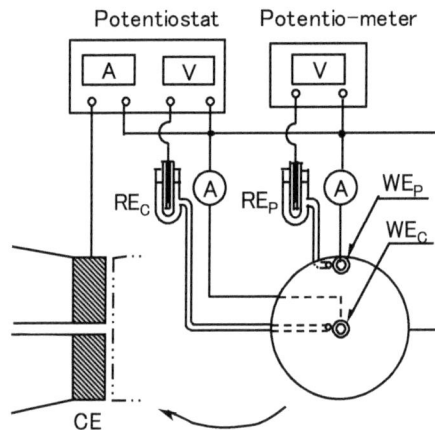

Fig. (4-37). Set-up for polarization curve measurement [3].

Secondly, he prepared an ingenious set-up for monitoring the polarization as shown in Fig. (**4-37**), with which external polarization curves were determined 20 min after the start of flow of the test solution (Fig. (**4-38**)). The relative positions of the polarization curves for the central part and the peripheral part coincide with those in Fig. (**4-34**), supporting the validity of the model in Fig. (**4-35**). In addition to this, it can be recognized that the polarization curves at the higher rate (0. 8 L min^{-1}) are located at the higher current side as compared with those the lower flow rate (0. 2 L min^{-1}). This explains the reason why the macro-cell current at larger flow rate was 200 times larger than that at the smaller flow rate.

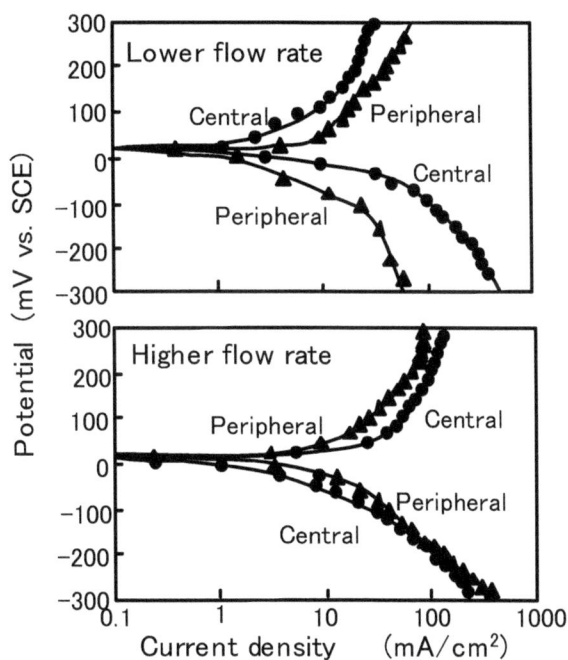

Fig. (4-38). Polarization curves obtained during corrosion tests in jet-in-slit with reverse flow at lower flow rate of 0. 2 L min^{-1} (upper) and 0. 8 L min^{-1} (lower) [3].

Lastly, after the reverse flow test at 0. 80 L min^{-1} he observed the micrograph of the specimen surface where the deep pit was originated: bare metal surface without any visible film was observed in the pit, whereas seemingly sufficient protective film was recognized on the circumferential surface, which also supported the model.

3. EROSION-CORROSION MECHANISM OF COPPER ALLOYS

For copper and copper base alloy, it has been clarified that one true colors of erosion-corrosion is a pure electrochemical corrosion which contains neither erosion component nor any influence of erosion at all. This was, however, already announced in the previous chapter. Newly added information in this chapter is that it is not a uniform corrosion but a localized corrosion which contains at least two anodic reactions with different dissolution rates within the system under consideration. The point is the agent which causes the difference in dissolution rate of anodic reaction. One of the agents is the flow velocity which is locally higher than that of the circumference. The shear force is accordingly high at the location of higher flow velocity, and it may separate the bulky but low strength oxide layer, with which such copper alloys as 6/4 yellow brass are characterized, exposing the under-laying metal surface to the fluid flow, hence accelerating the dissolution rate at the location. This is followed by the formation of a macro-cell of corrosion which consists of the macro-anode at the bare metal surface and the macro-cathode at the circumference where the dissolution rate is lower due to the oxide layer over the metal surface. Thereby the macro-cell corrosion process, namely localized corrosion advances at a high dissolution rate due to the macro-cell current effect as well as the surface area ratio effect, which cannot be achieved through a uniform corrosion.

Similarly with the shear force, the turbulence force originated in the flow through the deceleration of flow velocity also separates the protective layer from metal surface. Though, both forces function in the process of impingement attack the contribution of turbulence is by far larger than that of shear force. Inlet tube corrosion may be totally attributed to the turbulence force but not to shear force.

```
┌──────────────────────────────────────────────┐
│ Characteristic corrosion behavior such as with │
│ bulky but low strength oxide layer             │
└──────────────────────────────────────────────┘
        ┌────────────────────────────────────────────┐
        │ Irregularities in fluid flow over metal surface: │
     ◄──│ localized raise in shear stress and turbulence;  │
        │ localized lowering of flow velocity;             │
        │ generation of fixed vortex                       │
        └────────────────────────────────────────────┘
┌────────────────────────────────────────────────┐
│ Generation of difference in anodic dissolution rate of metal │
└────────────────────────────────────────────────┘
┌────────────────────────────────────────────────┐
│ Formation of macro-cell of corrosion accompanied with │
│ macro-cell current                               │
└────────────────────────────────────────────────┘
┌────────────────────────────────────────────────┐
│ Extension of difference in anodic dissolution rate through │
│ surface area ratio effect                        │
└────────────────────────────────────────────────┘
┌────────────────────────────────────────────────┐
│ Localized corrosion of high dissolution rate = Erosion-corrosion │
└────────────────────────────────────────────────┘
```

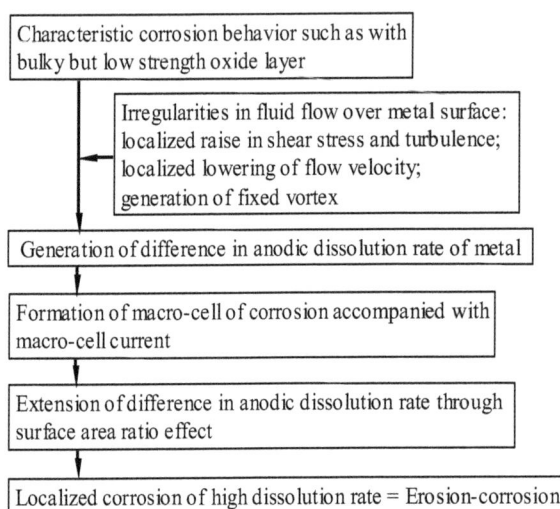

Fig. (4-39). Flow sheet describing the process of macro-cell formation for copper base alloys.

Another agent to cause the difference in dissolution rate of anodic reaction is lowering of local flow velocity. This arises more frequently due to a thermodynamic reason than the above mentioned rise in local flow velocity. Since the flow velocity is lower at the metal surface under the lump of dead water such as fixed vortex and hidden corner downstream of object, the supply of oxygen and anion is more meager than that at the opened circumference, and the oxide film which sufficiently demonstrates protective role is difficult to be formed. The dissolution rate from such metal surface is consequently higher than that from the circumference covered with the protective film. Thus, the difference arises at the anodic dissolution rate, and the macro-cell is formed in consequence, and the corrosion advances at high corrosion rate. It seems that most erosion-corrosion except for impingement attack arises through this mechanism. The description above is summarized in the flow sheet which is shown in Fig. (**4-39**).

REFERENCES

[1]   Matsumura M, Oka Y, Okumoto S, Furuya, H. Jet-in-Slit Test for Studying Erosion-Corrosion. In: Haynes GS, Baboian R, Eds. Laboratory Corrosion Tests and Standards; 1983: Bal Harbour USA: ASTM STP 866, Philadelphia, American Society for Testing and Materials, 1985; pp. 358-372.

[2]   Matsumura M, Oka Y, Yokohata H. Mechanism of Erosion-Corrosion on Copper Alloys. Boshoku-Gijutsu (presently Zairyo-to-Kankyo) 1986; 35: 706-711.

[3]   Murakami M, Yabuki A, Matsumura M. Corrosion of Pure Copper Caused by Vortex. Zairyo-to-Kankyo 2003; 52: 160-165.

[4]   Matsumura M, Yabuki A. Mechanism and Prevention of Erosion-Corrosion and Flow Velocity Difference Corrosion. Therm. Power 2005; 56: 192-203.

CHAPTER 5

Erosion-Corrosion Testing Methodology

Abstract: Using jet-in-slit apparatus, erosion-corrosion-proof test of 10 kinds of copper base alloys was conducted. Since a 1% copper chloride (II), $CuCl_2$, aqueous solution was adopted for the test liquid, corrosion rate of the test specimens was accelerated over 200 times higher than the rate of these materials in the field (as used for the potable water valves). Nevertheless, the ranking order of durability for the materials based on the test results agreed well with the ranking based on the experience of engineers who manufactured the water valves. Thus, owing to the materials comparison test, the reliability of jet-in-slit test methodology was confirmed. Further, a jet-in-slit apparatus was improved for conducting tests under the conditions of high temperature and high pressure which simulated the boiler feed water. With this jet-in-slit, corrosion tests were carried out on carbon steel and low alloy steel under various environmental conditions such as temperature, pH, and oxygen concentration as well as fluid flow conditions such as ordinary flow and reverse flow. Based on the test results the generation mechanism of erosion-corrosion on carbon steel in the water at elevated temperatures was clarified, which has made it possible to predict the erosion-corrosion damage on carbon steel under any environmental and fluid flow condition. Lastly, the mutually contradicting intentions of laboratory corrosion testing methodology were discussed: one is to obtain the test result in the shortest testing duration and another is to simulate most closely the corrosion mechanism of materials in the field.

Keywords: Jet-in-slit, copper base alloy, ranking order, carbon steel, boiler feed water, pure water, pH, dissolved oxygen concentration, erosion-corrosion, FAC.

1. JET-IN-SLIT TEST ON COPPER ALLOYS

1.1. Problems in Testing Methodology

Jet-in-slit may be ranked as one of the most excellent testing methods for erosion-corrosion. While the jet is small in scale and simple in structure, a number of flow patterns are produced on the specimen just by operating the flow direction and flow rate of the test liquid, and, in accordance with those flow patterns, various forms of erosion-corrosion are produced on the specimen. It is useful for demonstrating the essence of erosion-corrosion to be macro-cell corrosion. Nevertheless, the testing method is an instrument after all. Merits and demerits of the test equipment are dependent on the human ability which judges test result.

Most corrosion testing methodologies suffer from the dilemma mentioned in previous chapters: there is no qualification to be called testing equipment as long as the corrosion rate is not accelerated in it, but once it is accelerated it might not simulate the field corrosion any more. As a matter of course, the testing methodology which does not simulate the field is useless no matter how quickly the test result may be obtained. Then, the best policy may be to let it simulate the field at first, then to grope for the condition where the corrosion rate is accelerated as greatly as possible within the extent in which the corrosion mechanism is not hurt.

It could be a useful criterion for judging whether a testing method sufficiently simulates the field or not, to compare the ranking order of various materials in the testing equipment with that of the performance of materials in the field. In promoting this way, however, there is a serious problem in obtaining the ranking order of merit for the candidate materials in the field: it is almost impossible to manufacture real machines of equal type but of different materials and to derive them under equal operating conditions for longer duration of time, for example, a year at least. Thus, it is not simple even to evaluate the performance of various materials in the field.

1.2. Experience of Field Engineers

Domestic water taps are often made of copper alloys because of their sufficient durability in fresh water. Table **5-1** lists alloys that have been used or proposed for water taps in Japan. Except for the dezincification proof brass bars (DZPB), all other copper alloys are already registered in the Japanese Industrial Standards (JIS).

Table 5-1. Chemical compositions of copper base alloys for water tap use [1]

Material		Symbol	Composition (%)					
			Cu	**Zn**	**Pb**	**Sn**	**Fe**	**Ni**
Bronze	Bronze ingots for casting	BCIn 1	79 to 87	8 to 12	3 to 7	2 to 4	< 0. 35	< 0. 8
		BCIn 6	83 to 87	4 to 6	4 to 6	4 to 6	< 0. 3	< 0. 8
		BC 1	79 to 83	8 to 12	3 to 7	2 to 4	< 0. 35	< 1. 0
	Bronze castig	BC 1C	79 to 83	8 to 12	3 to 7	2 to 4	< 0. 35	< 1. 0
		BC 6C	83 to 87	4 to 6	4 to 6	4 to 6	< 0. 3	< 1. 0
Brass	Copper alloy	C3771BD	57 to 61	Bal.	1 to 2. 5	Sn + Fe <1. 0		-
	Rods and bars	C3604BD	57 to 61	Bal.	1. 8 to 3. 7	Sn + Fe <1. 2		-
	Brass castings	YBsC 3	58 to 64	30 to41	0. 5 to 3	< 1. 0	< 0. 8	< 1. 0
	Dezincification proof brass bars	DZPB I (0. 01As, Sb)	62. 4	Bal.	2. 4	0. 25	0. 17	-
		DZPB II (0. 01Sb)	62. 4	Bal.	1. 96	0. 28	0. 16	-

Some corrosion damage, however, was reported at valve seat rings, where the water flows with a relatively higher velocity (Fig. (**5-1**)). The possible influence of water flow on the damage of seat rings might result from cavitation erosion, or erosion-corrosion.

The former might promote localized damage through plastic deformation, which was originated by the attack of the impulsive high pressure of cavitation. The latter used to be regarded as the consequent of breakaway of the relatively thick layer of corrosion products by the action of flowing water, leading to corrosion damage with higher developing rate.

To cope with this problem, examination of the damage mechanism and the development of a test method for selecting durable materials were required. To examine the mechanism, damage similar to that in the field, had to be reproduced in the laboratory, which would generally take an impractically long duration of time. As to developing a new test method, it was doubtful whether a corrosion test conducted under an accelerated test condition rather different from that in the field would produce reliable results. However, no method other than an accelerated test would be of practical use.

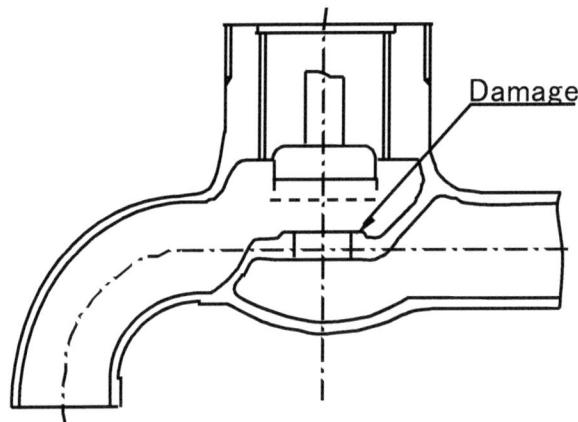

Fig. (5-1). Cross section of water tap: damage on the seat ring exposed to high-velocity flow of water [1].

To solve these mutually inconsistent problems at a stroke, the following research directions were determined: the candidate copper alloys (Table **5-1**) were to be examined using several test methods under the accelerated but well known experimental conditions: the damage mechanism in it had to be well-defined. Then, the alloys were to be ranked in order of merit according to results obtained in each test method. For different methods, the orders must be different, but some would coincide with the ranking order based on the experience of engineers in the field. Then, the damage mechanism in the field is to be regarded the same as that of the test method [1].

The field engineers experience relative to the damage of water taps was:

— bronzes were generally superior to brasses,

— among the bronzes in Table **5-1**, BCln 6 was superior to BCln 1,

— according to the engineers who manufactured hot water-supply systems, the brass C3771BD was superior to the brass C3604BD, and

— a water tap made of dezincification proof brass bar (DZPB) was rather sensitive to the damage at issue.

Three corrosion mechanisms were assumed for the damage of water tap seat: pure corrosion in the water at rest, the combined cavitation erosion and corrosion, erosion-corrosion where the corrosion products layer on the base metal might be separated through flowing liquid.

1.3. Accelerated Test Procedures

Pure Corrosion in Test Solution at Rest: A dip test conforming to ISO 6509 (Corrosion of Metals and Alloys — Determination of Dezincification Resistance of Brass) was adopted. A specimen with a test surface of 314 mm^2 was dipped in an aqueous solution of copper chloride (II) (12. 7 g $CuCl_2 \cdot 2H_2O$/1 000 mL H_2O, or 1% $CuCl_2$ solution). The container was sealed airtight and kept at 75 °C. After 24 h of dipping, the specimen was cleaned with acetone, and the mass loss of the specimen was determined by a balance to calculate the average corrosion rate.

Combined Cavitation Erosion and Corrosion: The vibratory cavitation test with an eccentric stationary specimen was adopted (Fig. (**5-2(a)**)). The vibrating nozzle was made of type 304 stainless steel and was of a columnar shape with a 16 mm diameter outside and a nozzle mouth of 1. 6 mm diameter inside.

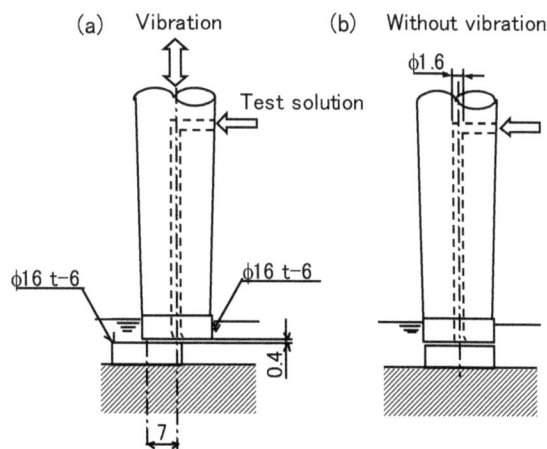

Fig. (5-2). Experimental devices: (a) vibratory cavitation unit with eccentric stationary specimen; (b) jet-in-slit apparatus [1].

It allowed injection of the test liquid (tap water, 40 °C, 0. 4 L min^{-1}) into the slit (0. 4 mm) between the nozzle and the specimen and vibration at a frequency of 20 kHz with the double amplitude of 25 μm to initiate cavitation on the surface and to move cavities in the specimen surface, resulting in cavitation damage. The nozzles and the specimens were of the same diameter (16 mm) but were arranged eccentrically 7 mm to each other so that a part of the specimen surface escaped damage. This area was the macro-cathode, and the damaged area served as the macro-anode, thus forming a macro-cell of corrosion in the specimen surface, which enhanced corrosion damage of the specimen.

Erosion-Corrosion: The jet-in-slit test was adopted in which 1% CuCl$_2$ solution was injected into the slit (0. 4 mm) between the specimen and nozzle of the same diameter (16 mm), as shown in Fig. (**5-2(b)**). The nozzle was made of polymethyl methacrylate resin with a bore diameter of 1. 6 mm. The flow rate of the test solution was 0. 4 L min^{-1} and the flow velocity at the nozzle outlet was 3. 3 m sec^{-1}. The solution filled up the slit and flowed radially over the specimen surface.

1.4. Test Results

Pure Corrosion in Test Solution at Rest: Macrographs of specimen surface after 24 h dipping are shown in Fig. (**5-3**). Dezincification was more or less recognized for three specimens of brass (C3604BD, C3771BD, and YBsC3), but not for others. The DZPB bars were also free from damage. The average corrosion rates in mm y^{-1} were determined for each specimen based on its mass loss during 24 h and are presented in Fig. (**5-4**). The test result agreed well with a field experience in that bronzes are superior to brasses in corrosion resistance. Some discrepancies, however, were found in the order of merit of BCIn 6 *vs.* BCIn 1, and C3604BD *vs.* C3771BD, as well as in DZPB I that exhibited nearly the same level of resistance as bronze. Accordingly, this test was judged not to have reproduced the tap water valve damage.

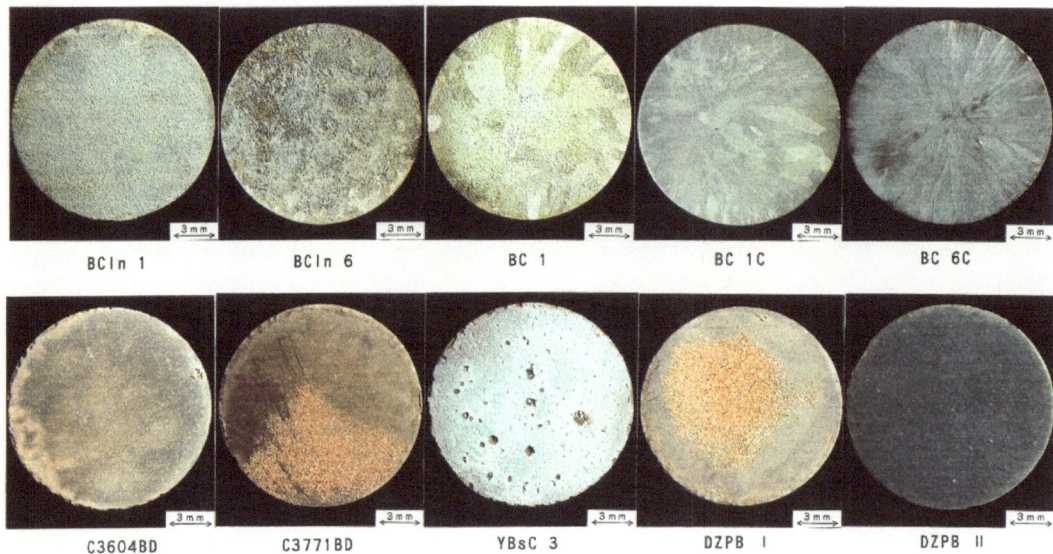

Fig. (5-3). Macrographs of test specimens after 24 h dip in a 1% copper chloride (CuCl$_2$) aqueous solution at 75 °C [1].

Combined Cavitation Erosion and Corrosion: The appearance of the damaged surfaces after the combined cavitation and corrosion test was quite similar. For example, a macrograph of the damaged BCIn 6 specimen surface is shown in Fig. (**5-5**).

Damage depth, *d* [μm], was adopted as the index of the damage. It was determined at a fixed point as the distance between the profiles of the original surface and that of the damage surface, measured with a surface roughness meter. The relationships between damage depth and testing time for specimens are shown in Fig. (**5-6**). Linear relationship between the depth and exposure time was observed for all copper alloys, although it was bent upwards for some of specimens: those casting materials (bronzes and YBsC3 brass) exhibited break points on their linear relationships of depth *vs.* exposure time, which made it difficult to judge precisely the order of merit. Nevertheless, a general survey of Fig. (**5-6**) indicated the superiority of brass over bronzes, which was contrary to field experience. Thus, this test was also judged to have failed in reproducing the valve seat damage of water tap.

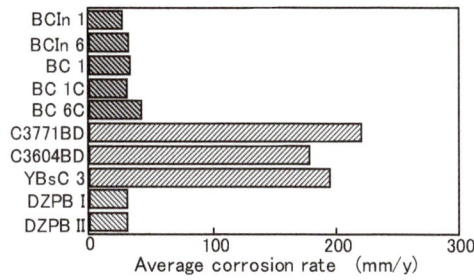

Fig. (5-4). Pure corrosion rates of copper alloys determined through mass losses during 24 h dipping in the test solution [1]

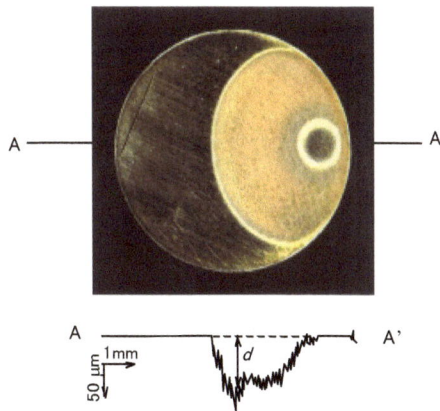

Fig. (5-5). Appearance of BCIn 6 specimen surface after 300 min exposure to cavitation (upper), and profile of the damaged surface (lower) [1].

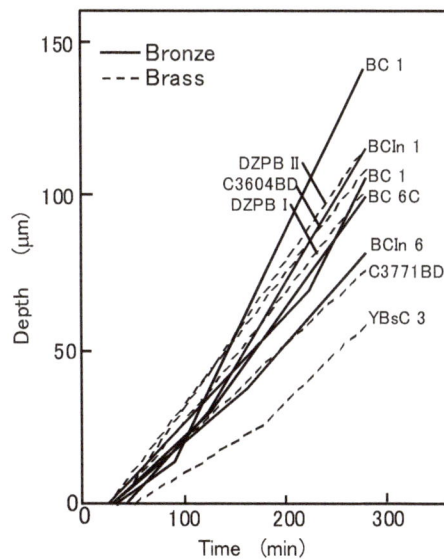

Fig. (5-6). Damage depth *vs.* testing time relationships for copper alloys obtained in vibratory cavitation test [1].

Erosion-Corrosion: Macrographs of specimens after a 420 min test in the jet-in-slit apparatus are shown in Fig. (5-7). Every surface of the bronze specimen was covered with a relatively thick layer of corrosion products. In contrast to this, dezincification (as identified by the change in color) as well as erosion-corrosion (as identified by the ring-shaped groove) occurred on the surface of the brass specimens. On the DZPB specimen surface, no dezincification occurred. But the ring shaped groove, that is, the damage caused by erosion-corrosion, was clearly noticeable. The plot of cumulative mass loss *vs.* exposure time for all 10 copper alloys is shown in Fig. (5-8), which could be categorized into two groups. The first one consisted of those curves of bronze that were convex upward, indicating that corrosion rate decreased with the exposure time. The second one was of brass specimens that were convex downward, indicating increasing corrosion rate with exposure time. Over the whole range of measurement, the bronze specimens exhibited smaller amounts of mass loss and lower corrosion rates than the brass specimens. The index in mass loss per damaged area showed a greater marked difference in corrosion rate. The superiority of bronzes over brasses in corrosion resistance was clear. In addition, as shown in Fig. (5-8), BCIn 6 was superior to BCIn 1, and C3771BD was superior to C3604BD. The DZPB 1 specimen showed a good resistance to dezincification but suffered the ring shaped damage indicating the possibility to localized corrosion of this sort. Since the above features of the test result were exactly consistent in all four aspects with the field experience, the damage on seat ring of water tap could be identified with erosion-corrosion. Accordingly, the jet-in-slit test using $CuCl_2$ solution was recommended for evaluation of material resistance.

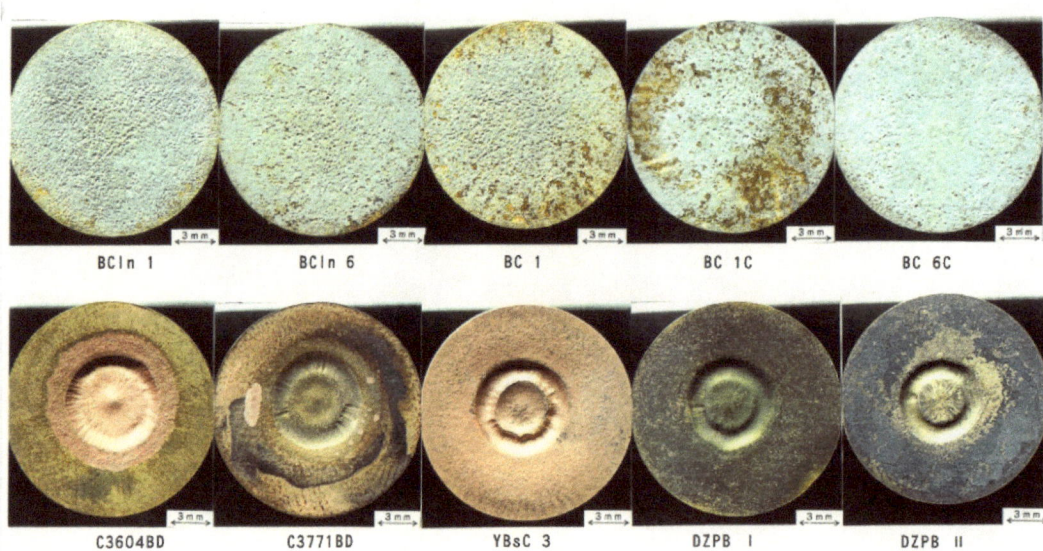

Fig. (5-7). Macrographs of specimen after 420 min in jet-in-slit test [1].

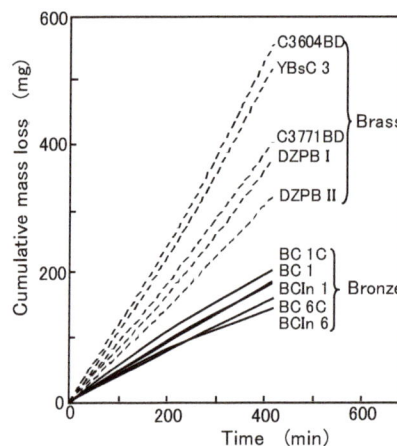

Fig. (5-8). Cumulative mass loss *vs.* testing time relationships for copper alloys obtained in jet-in-slit tests [1].

1.5. Function of Copper Chloride (CuCl$_2$) Solution in Accelerating Corrosion Rate

Function of Chloride Ion (Cl⁻): Fig. (**5-9**) shows the relationship between cumulative mass loss *vs.* test duration for a DZPB II specimen in jet-in-slit tests with various test liquids but of the same chloride (Cl⁻) ion concentration (0. 149 mol L^{-1}) at 40 °C: NaCl solution at pH 6, NaCl + hydrochloric acid (HCl) solution at pH 3. 5, and 1% CuCl$_2$ solution at pH 3. 5. The corrosion rate (slope of the lines) of the specimen in the 1% CuCl$_2$ solution was ca. 200 times higher than in the NaCl solution at pH 6. The addition of HCl to the NaCl solution of pH 6 to set its pH value at 3. 5 raised the corrosion rate about eight times higher, but it was still only one twenty fifth of that in the 1% CuCl$_2$ solution.

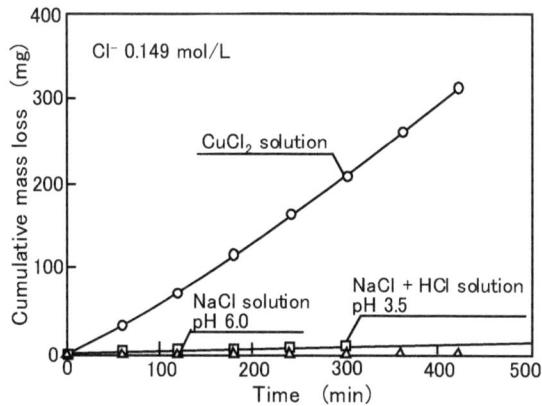

Fig. (5-9). Cumulative mass loss *vs.* test duration relationships for DZPB II specimen in jet-in-slit tests with various test liquids but the same Cl⁻ ion concentration at 40 °C [1].

Fig. (**5-10**) shows the effect of Cl⁻ ion concentration on the average corrosion rate of the DZPB II specimen in the jet-in-slit test. It was evident that anion concentration had similar influence on the corrosion rate in the CuCl$_2$ solution as in the NaCl solution and, accordingly, the anion concentration was not responsible for a larger difference in the corrosion rate between these test solutions. Thus, the acceleration of corrosion rate in the CuCl$_2$ solution was attributed neither to the anion nor to the pH value, but to the cation (i. e., Cu^{2+} ions in the solution).

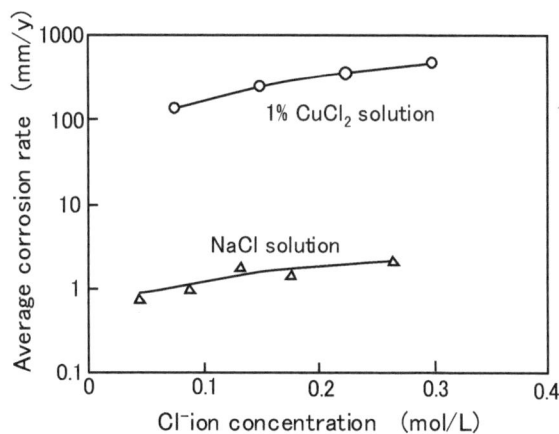

Fig. (5-10). Effect of Cl⁻ concentration on average corrosion rate for DZPB II specimen in jet-in-slit tests [1].

Function of Copper Iron (Cu^{2+}): The role of Cu^{2+} in accelerating the corrosion rate of copper alloys can be examined stoichiometrically as follows. When a brass that consists of α atoms copper (Cu) and β atoms zinc (Zn) is considered, then the corrosion reactions in a CuCl$_2$ solution will be:

$$\alpha \; Cu + \alpha \; Cu^{2+} \rightarrow 2\alpha \; Cu^{+} \tag{5-1}$$

$$\beta \; Zn + 2\beta \; Cu^{2+} \rightarrow \beta \; Zn^{2+} + 2\beta \; Cu^{+} \tag{5-2}$$

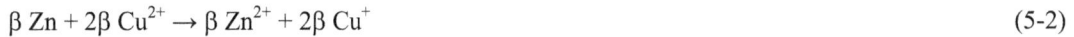

provided that the reduction of dissolved oxygen is neglected. Addition of Eqs. (5-1) and (5-2) gives:

$$\alpha \; Cu + \beta \; Zn + (\alpha + 2\beta) \; Cu^{2+} \rightarrow \beta \; Zn^{2+} + 2(\alpha + \beta) \; Cu^{+} \tag{5-3}$$

The term $2(\alpha + \beta) \; Cu^{+}$ corresponds to the increase in the amount of Cu^{+} ions in the test solution, and the sum of $\alpha \; Cu$ and $\beta \; Zn$ corresponds to the mass loss of the test specimen. Using 63. 5 as the atomic weight of Cu and 64. 5 as that of Zn, and 62. 4 as the atomic ratio of copper in the DZPB II specimen and 35. 2 as the ratio of zinc, the ratio (yield of Cu^{+} ion) *vs.* (the mass loss) is 1. 00 *vs.* 505.

In the case where dezincification occurs, Eq. (5-2) is replaced by:

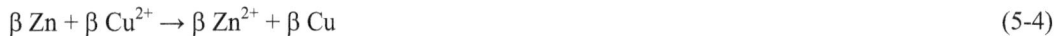

$$\beta \; Zn + \beta \; Cu^{2+} \rightarrow \beta \; Zn^{2+} + \beta \; Cu \tag{5-4}$$

Combining Eqs. (5-1) and (5-4):

$$(\alpha - \beta) \; Cu + \beta \; Zn + (\alpha + \beta) \; Cu^{2+} \rightarrow \beta \; Zn^{2+} + 2\alpha \; Cu^{+} \tag{5-5}$$

Using the preceding data again, the ratio (the yield of Cu^{+}) *vs.* (the mass loss) is 1. 00 *vs.* 0. 508.

A dip test was carried out on a DZPB II specimen in a 1% $CuCl_2$ solution (40 °C) for 420 min which resulted in a 26. 1 mg yield of Cu^{+} ions as determined by the thiocyanate method, and a 13. 9 mg mass loss of the specimen. The ratio was, accordingly, 1. 00 *vs.* 0. 532. This small difference in the level of ratio between this and that of preceding stoichiometric calculation was reasonably attributed to the contribution of oxygen. Thus, it was proven that Cu^{2+} ions played the role of oxidizing agent in the corrosion process of the copper alloys, and that most of the oxidized copper components dissolved into environmental solutions in the form of Cu^{+} ion.

1.6. Function of Fluid Flow

The dependence of the average corrosion rate of the specimen on the flow velocity of the test liquid, which was specified with that at the nozzle mouth, for those specimens of bronze (BCIn 1) and brasses (C3604BD and DZPB II) is given in Fig. (5-11). The corrosion rate of the bronze scarcely rose with the increasing flow velocity. In contrast, the corrosion rate of brasses suddenly jumped up at the velocities of 2. 5 and 3. 0 m sec^{-1} respectively. Observation on the specimen surface after the test runs with a flow velocity higher than the jumped-up velocity found that the surfaces of brass specimens consisted of a completely naked area and, at the same time, the other area still covered with the layer of corrosion products. In contrast, the bronze surface exposed to the same solution at equal velocity was found uniformly covered with a seemingly thick corrosion products layer as shown in Fig. (5-7). Thus, some phenomena which were similar to the breakaway velocity and erosion·corrosion which had been predicted by Syrett as illustrated in Fig. (1-11) were identified on the brass specimens but not on the bronze specimens.

A clear difference between erosion·corrosion (Fig. (1-11)) and erosion-corrosion (Fig. (5-11)) is that the former is a uniform corrosion, but the latter is a localized corrosion. Nevertheless, the characteristic velocity, no matter how it may be called breakaway velocity or jumping-up velocity, is a commonly important index in evaluating the anti erosion-corrosion durability of copper alloys in flowing environment since the relative erosion-corrosion rates of specimens determined at a flow velocity higher than the characteristic flow velocity may be different from those determined at a lower flow rate, even though the ranking order of the durability for them may not be changed.

We may conclude as follows. Just by allowing the test liquid to flow, the corrosion mechanism of the specimen was changed from uniform to localized, and by changing the flow velocity much more information was obtained for more accurate evaluation of test specimen.

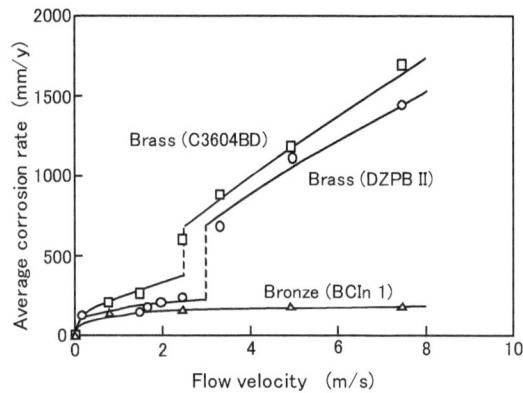

Fig. (5-11). Average corrosion rate *vs.* flow velocity relationships for brass specimens (C3604BD and DZPB II), and bronze (BCIn 1) in jet-in-slit tests [1].

2. LOCALIZED CORROSION ON CARBON STEEL IN FLOWING PURE WATER UNDER HIGH-TEMPERATURE AND HIGH-PRESSURE CONDITION

2.1. Jet-in-Slit Apparatus for Erosion-Corrosion Tests at Elevated Temperatures

Most of the boilers including the huge ones in thermal or nuclear power stations as well as those of the heat-recovery boilers in chemical factories are made of carbon steel. The feed water used in these boilers in Japan is mostly subjected to all-volatile treatment (AVT), a process that suppresses the concentrations of dissolved oxygen (DO) in the water to less than 7 ppb, causing a protective magnetite (Fe_3O_4) film to form on the surface of the carbon steel. At the same time, the water pH is controlled to around 9 in order to reduce magnetite solubility in the water, thereby stabilizing the protective magnetite. Nevertheless, unexpected occurrences of severe localized wall thinning of carbon steel pipes and resultant failures have been reported: anomalously fast (> 1 mm y^{-1}) progress of localized corrosion in places where the water flow might be either turbulent or too fast. These phenomena, identified as flow-induced localized corrosion or flow accelerated corrosion (FAC), must be at least a form of erosion-corrosion.

One of the possible ways to cope with this problem is to reproduce the corrosion in laboratory and elucidate in detail the corrosion mechanism. To do this, a jet-in-slit apparatus for high temperature use was developed by Tachibana, an excellent graduate student of Hiroshima University. Fig. (5-12) is a schematic rendering of the jet-in-slit apparatus constructed by him for testing in high-temperature and accordingly high-pressure water [2].

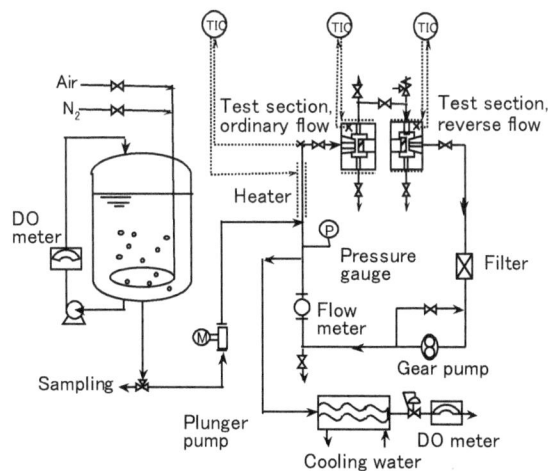

Fig. (5-12). Jet-in-slit arrangement for corrosion test under high pressure at elevated temperatures [2].

The main components were a water-conditioning tank, a test-solution feed line, a test-solution circulation line, and a test-solution drain line. A pair of test cells was installed in the test-solution circulation loop. After its quality adjustment, the feed water was pressurized with a plunger pump and sent into the circulation loop at the flow rate of 500 mL h^{-1}, and the amount of water equal to that of feed water was drained from the loop through a back-pressure valve. The maximum test temperature in the loop was 200 °C and 10 MPa.

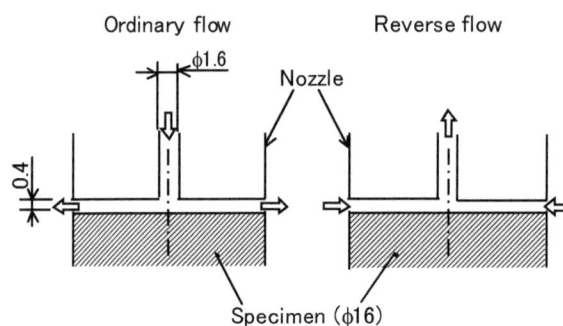

Fig. (5-13). Flow direction on specimen surface [2].

The test section installed in the test-solution circulation line was a set of jet-in-slit cells for ordinary flow and reverse flow. In Fig. **(5-13)**, the specimens are depicted under the nozzles in order to clearly present the difference in flow condition: in the ordinary flow configuration, a disk-shaped test specimen 16 mm in diameter was placed in front of a nozzle 1. 6 mm in mouth-opening diameter, 0. 4 mm in gap distance. Water from the nozzle was allowed to flow radially on the specimen surface decelerating its velocity due to the larger cross-section of flow as it approached the periphery of specimen. As a consequence of the rapid deceleration of flow velocity an intense turbulence occurred in the water flow over the specimen surface; localized corrosion due to the shear force and turbulence force were originated on the copper alloy specimen as described in the previous section. In contrast to this, no turbulence was generated in the reverse flow at all, since the flow velocity was not decelerated but accelerated as the water approached from the periphery to the center of specimen. Localized corrosion due to the flow velocity difference as well as that due to fixed vortex occurred, as also described in the previous chapter, on the specimen surface of copper alloy. The specimens were fixed through a Teflon bush to ensure electrical insulation from the test cell.

2.2. Treatments for Boiler Feed Water

In European countries, combined water treatment (CWT) and neutral water treatment (NWT) are used more widely than AVT. In both of these processes, DO (dissolved oxygen content) is maintained at 20-200 ppb, controlling the water pH in a range 6. 5-9. 0, the purpose of which is believed to make the protective surface film dense by filling the pores of initially formed magnetite (Fe_3O_4) film with fine particles of hematite (Fe_2O_3) that have relatively low solubility in water. Thanks to this processing, the surface film is smoothed, providing the additional advantage of reducing pressure loss. This advantage has led to examining the introduction of CWT in some boilers in Japan.

With the ultimate aim to identify the factors causing the anomalous corrosion of carbon steel in boilers, the authors first set out to clear up the influence of water treatment on corrosion, that is, to conduct experiments in three different water conditions which are subjected to AVT, CWT and NWT using the jet-in-slit to generate the conditions which would lead to the various forms of erosion-corrosion.

The test specimen was of JIS S25C carbon steel (0. 25%C, 0. 25Si, 0. 45Mn, 0. 030P, S). Before the test, the surface of the specimen was emery-polished down to #2000, then washed and degreased. The test solution was ion-exchanged water with electrical conductivity of 1×10^{-5} S m^{-1}. To simulate industrial boiler feed water conditions, the DO and pH of the test solution were set as indicated in Table **5-2**.

Table 5-2. DO and pH in the water of various treatments [2]

	AVT	CWT	NWT
DO (ppb)	<10	100	100
pH	9	9	7
Temperature (°C)	80-200		
Pressure (MPa)	10		
Flow velocity (m/s)	0-6. 6		

The DO was maintained at a specified level by continuously monitoring the content in the drain water and controlling the amount of nitrogen and air injected into the tank for water quality adjustment: the DO level of the water in the tank was set a bit higher than the specific level to compensate for the oxygen consumed by the corrosion reaction of the specimen. To adjust the pH, ammonia was injected. The temperature of test water ranged between 80-200 °C, pressure 10 MPa, and flow velocity at the nozzle exit 0-6. 6 m sec^{-1}.

Dependence of Mass Loss on Test Duration: Fig. (**5-14**) plots the observed mass loss of specimen as a function of testing time for 17 specimens in NWT water at 120 °C and 10 MPa, and flow velocity at nozzle exit being 3. 3 m sec^{-1}. The mass loss that jumped up at 5 h was rather modest from 5 to 15 h and steady from 20 to 50 h. Based on the judgment that the corrosion reaction reached the steady state after 40 h, it was made the standard test duration in the subsequent experiments.

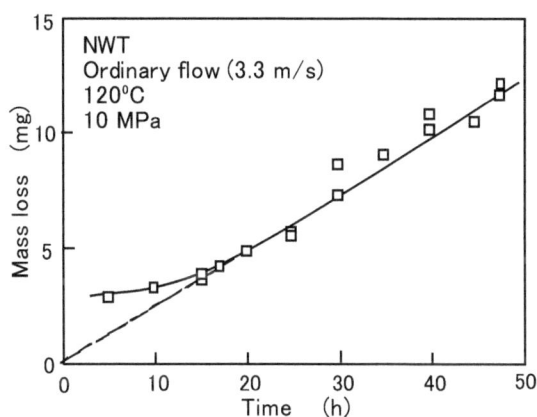

Fig. (5-14). Increase in mass loss with the duration of test for carbon steel specimen [2].

Influence of Water Treatments: The first set of experiments, with water of three different treatments, was conducted for 40 h at 120 °C and 10 MPa, with various flow velocities using the ordinary flow cell as well as the reverse flow cell at the same time. Fig. (**5-15(a)**) summarized the results obtained in the ordinary flow cell. The 10 mg mass loss at 40 h corresponds to an average corrosion rate of 1. 4 mm y^{-1} as it is indicated in the vertical axis right side. In AVT and CWT water, the mass losses under flowing water were similar to those in the stagnant conditions but those in NWT water were considerably higher than in the stagnant NWT water, and when compared at the same flow velocity, the mass loss in flowing NWT water was considerably greater than that in AVT or CWT water. The trends in the reverse flow shown in Fig. (**5-15(b)**) resembled those in the ordinary flow.

The macrograph in Fig. (**5-16**) compares the appearance of specimens tested in the three different water conditions, AVT, CWT and NWT, in the ordinary flow at 3. 3 m sec^{-1} and 120 °C after the removal of oxide layer from the surface. The corresponding cross-sectional profiles are also given. A ring-shaped attacked zone was observed on the surface of the AVT specimen. The occurrence location of the attacked zone coincided well with the area of turbulence identified in the previous chapter.

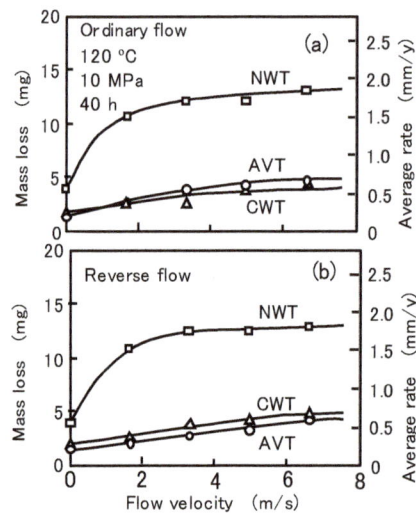

Fig. (5-15). Effect of flow velocity on the mass loss of carbon steel specimen in various boiler feed water environments at 120 °C: test duration, 40 h; (a) ordinary flow; (b) reverse flow [2].

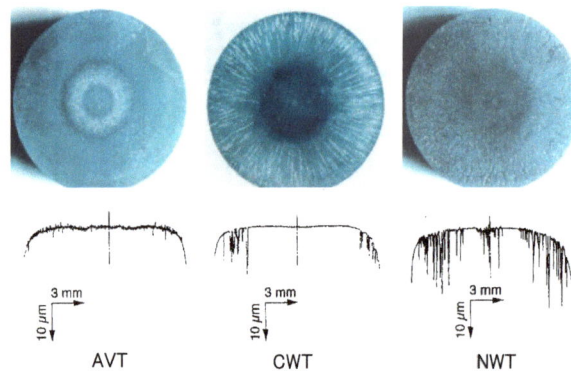

Fig. (5-16). Observation of surface and cross section of carbon steel specimen after corrosion test of 40 h in various boiler water environments at 120 °C [2].

The corresponding cross-sectional profile indicated that the zone was only around 1 μm deep. As to the CWT specimen, the central part of specimen surface remained intact and retained its metal luster, and the cross-sectional profile in the part was smooth. In contrast, in the periphery part, relatively deep grooves (5-10 μm) developed radially. They might look like deep pits in the cross-sectional profile, but they were actually grooves as it was indicated in the surface observation. The entire surface of the NWT specimen was rough, and radially developed groove just like the ones on the CWT specimen were also observed in the periphery. The depth of grooves on the NWT specimen was 5-20 μm, being larger than those on CWT specimen.

Since the distribution of damage was not uniform on the surface of each specimen (AVT, CWT and NWT) as it could be seen in Fig. (5-16), mass loss was revealed not a satisfactory index for characterizing the concerned damage. It was decided, therefore, to take the following indices as measures of corrosion damage: maximum depth of corrosion at 2 mm from the center for the AVT specimen, and depth of the deepest groove near the edge for the CWT and NWT specimens.

Fig. (**5-17**) summarized the maximum depth of corrosion observed on the specimens after 40 h exposure in 120 °C water at 10 MPa: (a) ordinary flow and (b) reverse flow. A maximum depth of 5 μm in these plots corresponded to a maximum rate of penetration of 1. 1 mm y^{-1}.

Comparing AVT and CWT in terms of maximum depth, the corrosion in CWT water was 3-5 times greater than that in AVT water, in spite of that they were rather equal in terms of mass loss (Fig. (**5-15**)). Represented in mass loss or maximum depth, difference in corrosion damage between the stagnant water and flowing water was negligible for AVT but significant for CWT and NWT, and also with no appreciable difference in corrosion performance between the ordinary flow and the reverse flow. Views for each result are as follows.

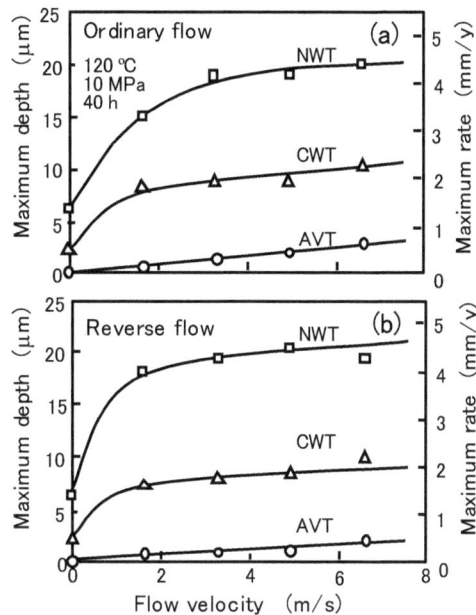

Fig. (5-17). Effect of flow velocity on the maximum depth of corrosion caused to carbon steel specimen in water of various boiler feed environments: (a) ordinary flow; (b) reverse flow [2].

In AVT water, the ring-shaped damage emerged on the specimen in the same place where similar damage had occurred on the copper alloy specimens as described in Chapter 4. At a flow velocity of 3. 3 m sec^{-1}, the maximum depth of corrosion on carbon steel was relatively shallow, 1 μm per 40 h or 0. 2 mm y^{-1}. The difference in amount of damage between the stagnant water and flowing water was rather small. The breakaway velocity at which the protective film of copper based alloy was used to separate from the metal surface was not observed. Moreover, a similar damage behavior was observed irrespective of flow direction: ordinary flow or reverse flow. Thus, for AVT, the flow of water had a trivial effect on the damage, although some slight influence of erosion-corrosion might have been observed.

In CWT water, the specimen surface remained smooth and retained its original metallic luster, showing no evidence of thinning at the location of high shear force and intense turbulence, that is, the central part of the specimen surface. This supported the claim that in actual boilers pressure loss was smaller with CWT water than with AVT water. However, in the periphery of the disk specimen surface, distinctive groove corrosion (> 1 mm y^{-1}) developed, and the maximum depth of corrosion was significantly influenced by the water flow (Fig. (**5-17**)). This suggested that water flow had a significant effect on the damage of CWT specimen, even though it might be a different form of erosion-corrosion.

The NWT specimen was damaged, in both mass loss and maximum depth, more than twice severely as compared to the CWT specimen. The cross-sectional profile showed an evidence of attack near the center of specimen, the location where no damage developed for the CWT specimen.

Influence of Water Temperature: The AVT and CWT were further compared on the effect of water temperature on the damage. Since there was no significant difference in the results obtained in ordinary or reverse flow, only the results obtained in ordinary flow are presented here. Also, since the damage behaviors of NWT and CWT specimens were not very different, the result of NWT will be not discussed. Fig. **(5-18)** plots the observed mass loss after 40 h test as a function of water temperature for AVT and CWT specimens. The maximum mass loss was realized at 160 °C for both AVT and CWT. Fig. **(5-19)** plots the observed maximum depth of corrosion after 40 h as a function of the water temperature for the same specimens. An unexpected result was that the temperature yielding the peak depth for CWT (120 °C) was considerably lower than that for AVT (160 °C). It was also lower than those temperatures which yielded the peak mass loss for both AVT and CWT (160 °C).

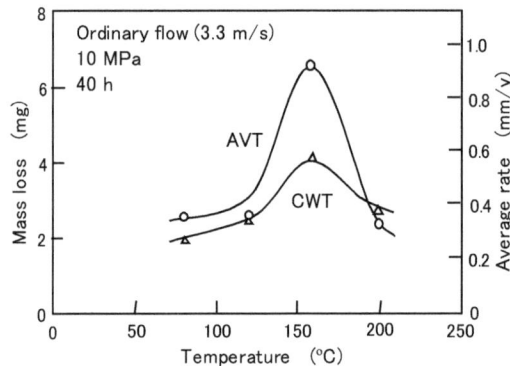

Fig. (5-18). Effect of temperature on mass loss of carbon steel specimen; test duration, 40 h [2].

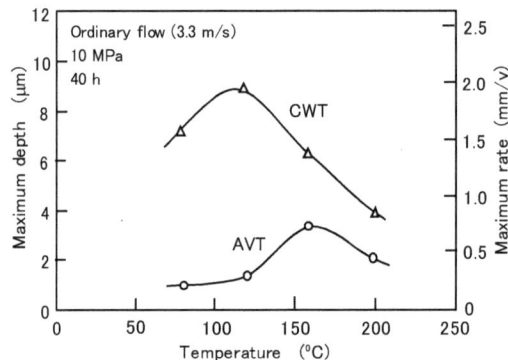

Fig. (5-19). Effect of temperature on maximum depth of corrosion originated in carbon steel specimen [2].

Composition of Oxide Film: One of the significant factors influencing the concerned corrosion morphology would be the quality of the oxide film deposited on the specimen surface. According to the literature, the composition of oxide film is reflected in the corrosion potential of specimen. During the jet-in-slit test at elevated temperature, the corrosion potential of specimen was monitored in the ordinary flow cell. The observed variation patterns of potential with time are shown in Fig. **(5-20)**: around five hours after the beginning of the test, corrosion potentials reached steady level; rather lower potential for AVT specimen (-600 mV), and higher for CWT and NWT specimens (-150 mV).

The distinct differences in the specimen potential between the former and latter two might be attributed to the ten times higher DO for CWT and NWT than that for AVT. According to the available potential *vs.* pH diagram [3], magnetite, Fe_3O_4, was the most stable compound under the AVT condition, and hematite, Fe_2O_3 under the CWT and NWT conditions. Visual inspection showed that the AVT specimen was totally covered with dark film, whereas the CWT and NWT specimens were partially covered with brown as well as dark film. These findings confirmed that the AVT specimen was completely covered with magnetite film, whereas on the CWT and NWT specimens, both hematite and magnetite films were present.

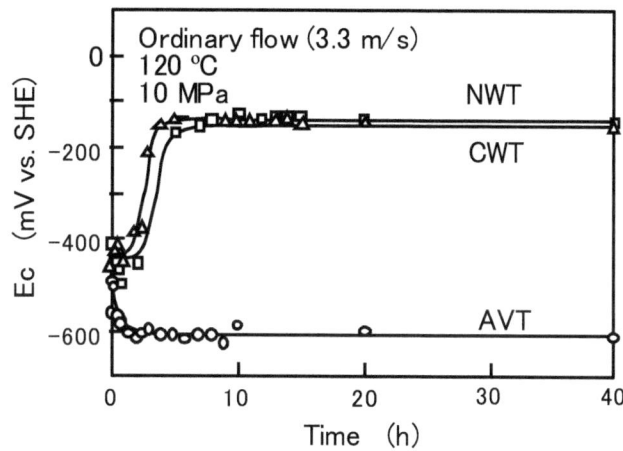

Fig. (5-20). Corrosion potential of carbon steel in water of various boiler feed environments [2].

A small conclusion here would be as follows: Rising of the dissolved oxygen content (DO) in boiler feed water affects not only the composition of the corrosion product layer on the carbon steel but also the corrosion morphology in the way of receiving the effect of fluid flow. As long as it is compared in the maximum depth of corrosion, AVT water with the lower DO is seemingly advantageous in preventing corrosion over CWT or NWT water which contains ten times more oxygen. We cannot, however, attribute the advantageous performance of AVT in preventing the corrosion damage solely to its DO, since there were other influencing factors which must be taken into consideration.

2.3. AVT Water with Various pH Levels

As described above, AVT water with the lower DO was seemingly advantageous in preventing corrosion, as long as it is compared in maximum corrosion depth, over CWT and, in particular, over NWT water. This might be naturally attributed to the DO in the water. Another possibility might be the pH of water, since that of NWT water was lowest but the corrosion damage was largest among the three kinds of water treatment. The corrosion damage in AVT water, therefore, may be further reduced, when pH of the water is raised. Effect of pH level of AVT water on the corrosion damage to carbon steel in it was investigated under the conditions described in Table **5-3** using the same apparatus and with the same procedure that was described in the previous section [4].

Table 5-3. Test conditions in AVT water of various level of pH [4]

Test material	Carbon steel (S25C)
Test duration (h)	40
Temperature (°C)	120
Pressure (MPa)	10
Flow velocity (m/s)	3. 3
DO (ppb)	<10
pH	9. 0-10. 0

Damage Morphology: Fig. (**5-21**) shows the sketch of specimen surface appearances after 40 h tests. The corresponding surface cross-sectional profiles (after removal of the surface oxide film) are given in Fig. (**5-22**). As it can be seen in this figure, the broad but shallow damage (indicated by arrows for the specimens at pH 9. 0) as well as the narrow but deep damage (indicated by arrows for the specimen at pH 9. 5) were observed. Since the integrity of a boiler system in service depends on the deepest damage, we decided to evaluate corrosion performance in terms of the maximum depth of corrosion again.

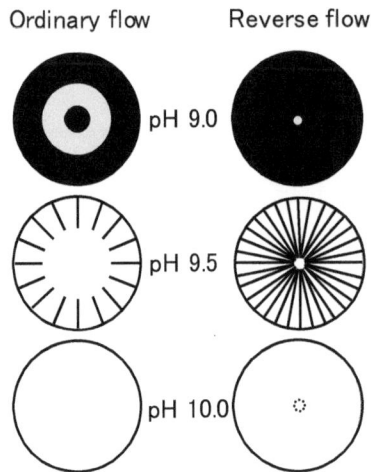

Fig. (5-21). Sketches showing corrosion morphology on carbon steel specimen obtained in various pH environments: temperature, 120 °C [4].

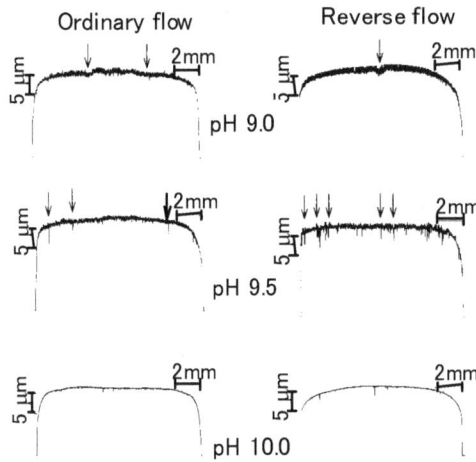

Fig. (5-22). Cross-section of carbon steel specimen after 40 h corrosion test at various pH at 120 °C [4].

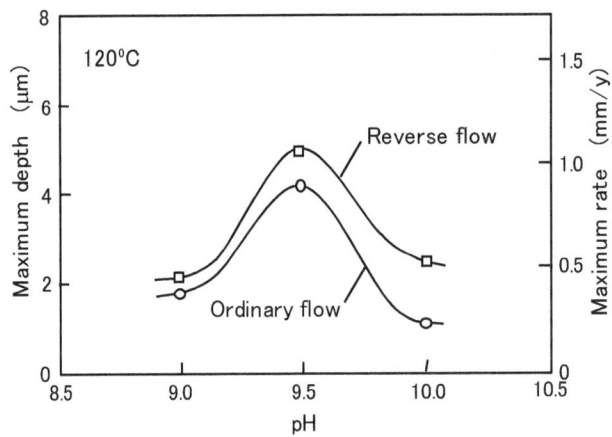

Fig. (5-23). Effect of pH on the maximum depth of corrosion for carbon steel specimens: temperature, 120 °C; duration of test, 40 h [4].

Fig. (**5-23**) plots the observed maximum depth of corrosion (μm) against pH of AVT water. The right hand vertical axis refers to that converted into maximum rate of penetration (mm y^{-1}). The maximum rate of penetration in the water at pH 9. 0 was around 0. 4 mm y^{-1} for both ordinary flow and reverse flow. In the water at pH 9. 5, however, it rose rapidly up to 0. 9 (ordinary flow) and 1. 1 (reverse flow). Then, at pH 10. 0, it rapidly dropped down to 0. 2 (ordinary flow) and 0. 5 (reverse flow) mm y^{-1}. A similar rapid up and down was recognized in the pH dependence of mass loss as shown in Fig. (**5-24**). Since the pH dependence of corrosion rate was so intense, all of the measurements were repeated three times under the same conditions, thus the values plotted in Figs. (**5-23**) and (**5-24**) are the average of each set of three.

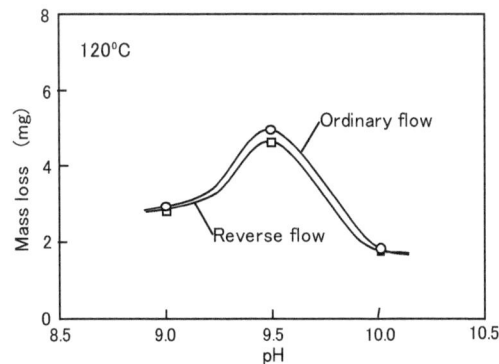

Fig. (5-24). Effect of pH on mass loss of carbon steel specimens: temperature, 120 °C; duration of test, 40 h [4].

According to the careful inspection of characteristic features of damage morphologies, as presented schematically in Fig. (**5-21**), the surface of specimen at pH 9. 0 was covered with a black film irrespective of the flow direction of water, that is, ordinary flow or reverse flow. The gray intermediate ring on the specimen tested under ordinary flow at pH 9. 0 corresponded to the broad but shallow attacked area indicated by the arrows in Fig. (**5-22**). At pH 10. 0, the surface of the specimen tested under ordinary flow completely retained its metallic luster, except for slight spot just beneath the nozzle in reverse flow. At pH 9. 5, its metal luster retained on the surface except for the radial linear damage, that is, the groove already referred in Fig. (**5-16**). The line density and length of the groove were greater in the reverse flow than in the ordinary flow.

Corrosion Potential: Fig. (**5-25**) summarizes the observed variation patterns of corrosion potential under the ordinary flow. At every examined pH, corrosion potential varied only in the initial 5 h period and then stabilized at its respective steady state level. The steady state potential level was the least noble at pH 9. 0, yielding -550 mV, and was around -350 mV at pH 9. 5 and 10. 0. According to the potential-pH diagram, magnetite (Fe_3O_4) must be the most stable phase at pH 9. 0, and hematite (Fe_2O_3) at pH 9. 5 and 10. 0.

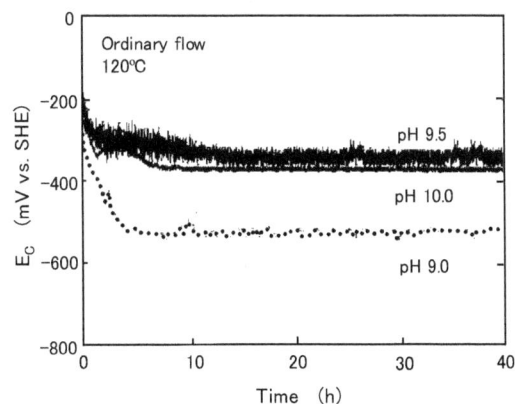

Fig. (5-25). Corrosion potential for carbon steel specimen in various pH environments [4].

Thus, the black film on the specimen tested in the water at pH 9. 0 and black lines at pH 9. 5 was identified as magnetite, while a protective oxide film like $\alpha\text{-}Fe_2O_3$ and $\gamma\text{-}Fe_2O_3$ must have been present on the surface with the metallic luster of the specimen tested at pH 9. 5 and 10. 0.

2.4. Effect of Alloy Element in Preventing the Corrosion Damage

In this section, we evaluate the performance of low alloy steel, which had already been used in an actual boiler as a substitution of the carbon steel pipe in which corrosion damage of issue was generated (the detail will be discussed in Chapter 7). It is also under consideration as a structural component in the concerned corrosion environment for boilers under plan.

A low alloy steel, JIS SCM 415, containing 1% chromium (0. 15C, 0. 25Si, 0. 75Mn, <0. 03P, S, 1. 0Cr, 0. 2Mo) was used as the test specimen. Prior to the test, the specimen surface was emery-polished down to #2000 and washed and degreased. The test solution was ion exchange water with an electric conductivity of 1×10^{-5} S m^{-1}. To simulate the industrial boiler water quality, DO was controlled to be less than 10 ppb and pH was set in the range of 9. 0-10. 0. Ammonia was injected for pH control of the test water. The temperature of the water varied in the range of 80-200 °C under the setting of pressure 10 MPa, and flow velocity at the nozzle mouth was maintained at 3. 3 m sec^{-1} during 40 h of testing. The amount of corrosion was evaluated with the mass loss which was determined after removing the surface oxide film by electrolytic polishing. The surface profile was determined by using a surface roughness-meter in order to observe the development of localized corrosion [5].

Influence of pH: Fig. (**5-26**) summarizes the macrographs of representative specimen surfaces after 40 h corrosion test in the jet-in-slit apparatus at 120 °C under varying pH conditions. The corresponding cross-sectional profiles of the specimen surface are presented in Fig. (**5-27**). Under the condition of ordinary flow of water at pH 9. 0 and reverse flow at pH 9. 0 and 9. 5, radial lines emerged on the specimen surface (Fig. (**5-26**)), and they were identified as the narrow but deep grooves of corrosion.

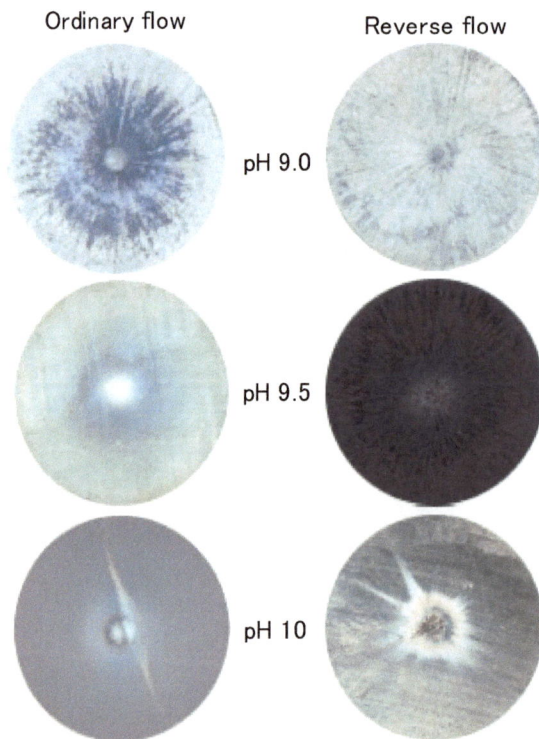

Fig. (5-26). Morphology of low alloy steel specimen surface after a 40 h corrosion test at various pH of water: temperature, 120 °C [5].

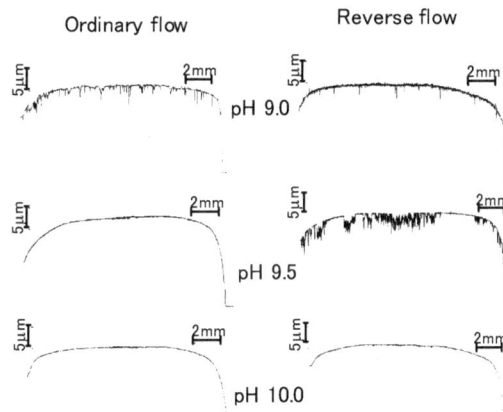

Fig. (5-27). Cross-section of low alloy steel specimens after a 40 h corrosion test at various pH of water: temperature, 120 °C [5].

The number of grooves was especially large under the condition of reverse flow at pH 9. 5. On the other hand, the cross-sectional profiles showed generally smooth surface, regardless of the center part or the periphery of the specimen, for those specimens tested in ordinary flow at pH 9. 5 as well as at pH 10, and reverse flow at pH 10 in spite of some spots that appeared on the specimen surface.

For the boiler service in the field, the concerned factor is the deepest corrosion damage, and accordingly the test results were evaluated in terms of the deepest corrosion detected for the specimen. Fig. (5-28) plots the observed maximum depth of corrosion for the low alloy steel specimen in the water at 120 °C as a function of pH. The scale given in the right hand side vertical axis refers to the maximum rate of penetration. Both for ordinary flow and reverse flow, the maximum rate of penetration reached as high as 1 mm y^{-1} with pH 9. 0, showing a trend of decreasing with further rise in pH.

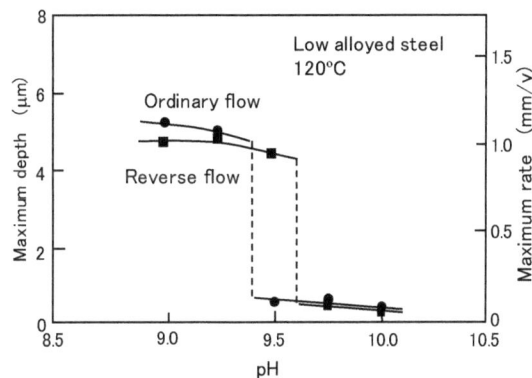

Fig. (5-28). Effect of pH on the maximum depth of corrosion on low alloy steel specimens: temperature, 120 °C; test duration, 40 h [5].

Discontinuous sudden decrease in the maximum depth of corrosion was detected in the range of pH 9. 25-9. 50 for ordinary flow, and in the range of pH 9. 5-9. 75 for reverse flow. As a result, the damage depth at pH 9. 5 was quite different between ordinary flow and reverse flow.

The observed discontinuous sudden decrease in the maximum depth of corrosion in the above-mentioned pH ranges must be the consequence of the uniform formation of the protective film over the steel surface at the pH that exceeded a threshold level. The difference in the corrosion depth sudden-decreasing-pH range between the ordinary flow and the reverse flow condition must be ascribed to the difference in flow state between ordinary flow and reverse flow (detail will be discussed later). The plot of mass loss against pH (Fig. (5-29)) was compatible with that of maximum depth against pH.

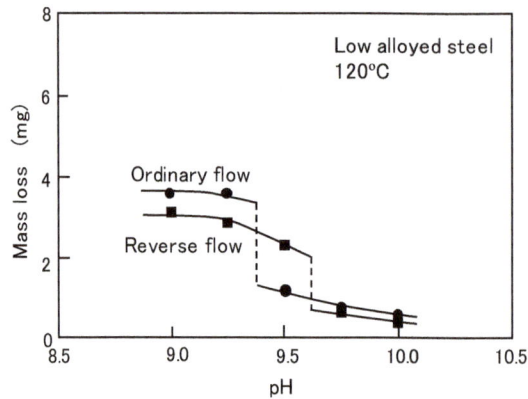

Fig. (5-29). Effect of pH on mass loss of low alloy steel specimens: temperature, 120 °C; test duration, 40 h [5].

Influence of Temperature: Fig. (5-30) summarized the maximum depth of corrosion observed in the water of pH 9. 0 as a function of temperature. For the sake of comparison, the data obtained for carbon steel were also plotted. On the right hand vertical axis, the maximum rate of penetration was given. Under either ordinary flow or reverse flow condition, the maximum depth of corrosion tended to rise with the temperature up to 120 °C to yield a constant level over the range of 120-160 °C. This steady-state level of the maximum depth for the low alloy steel was 1. 1 (reverse flow) and 1. 2 mm y^{-1} (ordinary flow), which was by two times greater than that of the carbon steel. Between 160 °C and 180 °C, the discrete drop in maximum depth down to 0. 2 mm y^{-1} was observed for the low alloy steel, which was much lower than that for the carbon steel. This difference may be attributed to the superior corrosion resistance of the passive film deposited over the low alloy steel surface in this range of temperature over that of carbon steel.

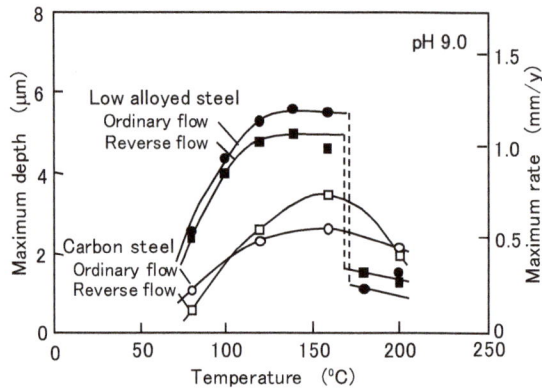

Fig. (5-30). Effect of temperature on the maximum depth of corrosion for low alloy steel and carbon steel specimens: test duration, 40 h; pH 9. 0 [5].

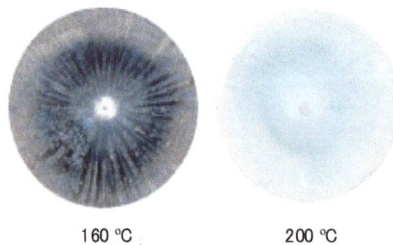

160 °C   200 °C

Fig. (5-31). Specimen surfaces after a 40 h corrosion test at pH 9. 0 at 160 °C and 200 °C [5].

Fig. (**5-31**) compares the macrographs of specimen surface after a 40 h test under the reverse flow condition at 160 °C and at 200 °C. Corrosion groove was observed on the specimen surface tested at 160 °C, but the surface remained smooth at 200 °C, showing a similar variation of corrosion pattern with respect to pH (Fig. (**5-26**)).

In short, this evidence indicated that, at temperatures higher than 180 °C, corrosion resistance of the low alloy steel was superior to that of the carbon steel, but, at temperatures lower than 160 °C, the groove corrosion developed on the low alloy steel was deeper than that on the carbon steel. Thus, acceptance of low alloy steel as a component of boiler under plan may be favorable if the boiler operation condition is to yield formation of homogeneous protective surface layer, but it must be avoided when the boiler is to be operated under a condition where groove corrosion yields on it.

2.5. Mechanism of Corrosion Groove Formation

For operating boilers safely, the maximum depth of corrosion is, as repeatedly pointed out, one of the highly concerned factors. In the jet-in-slit tests conducted on carbon steel and low alloy steel so far, the maximum depth of corrosion was brought about by groove corrosion. Consequently, the generation mechanism of groove corrosion becomes one of the largest interests. As it can be seen in Figs (**5-26**) and (**5-27**), groove corrosion developed at 120 °C in the water with pH 9. 5 under the reverse flow but not under the ordinary flow at all. This clear distinction appears to indicate that the development of groove corrosion must be critically influenced by the water flow condition: the grooves formed along the flow direction from the periphery to the center of specimen; the site of incidence of grooves must obviously be in the very close vicinity of the specimen edge.

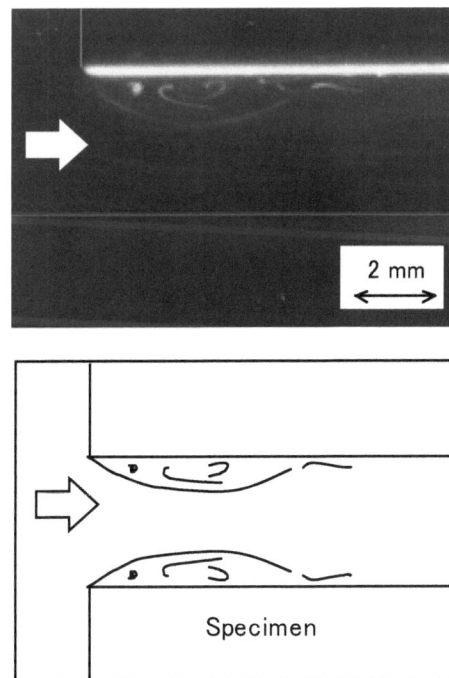

Fig. (5-32). Observation of vortex-string in fluid flow near a step (upper) and the flow pattern expected near the actual edge of specimen (lower) [5].

To look into this aspect, the flow pattern near the edge of specimen was visualized under the reverse flow condition. Fig. (**5-32**) shows the observed flow visualization with a flow velocity of 1. 3 m sec^{-1}. The white line in the figure refers to the stream line of the tracer; the higher the flow velocity, the longer would be the stream line or *vice versa*. It could be seen in the figure that a vortex was induced at the very edge, and the flow velocity in the vortex was only one third of the water outside. This must be that sort of vortex which is called vortex-string or vortex rod in fluid dynamics. In the actual experimental apparatus of jet-in-slit, the specimen edge coincided with the nozzle edge as depicted at the bottom sketch in Fig. (**5-32**). The vortex of this mode is supposed to arise from the edge of the specimen, and then develop downstream on the specimen surface to reach its center.

On the other hand, as to the ordinary flow specimen with the groove corrosion, for example, the carbon steel specimen tested in CWT water (Fig. (**5-16**)) or that in AVT water of pH 9. 5 at 120 °C (Fig. (**5-22**)), a thin but stable surface film composed mainly of hematite (Fe_2O_3) must have been formed in the central part of the surface where water was flowing fast resulting in the relatively enhanced rate of oxygen supply rendering the central part remained virtually intact. In contrast, near the edge of the disk specimen, where the flow velocity was relatively low, discrete rust cob composed mainly of voluminous magnetite must have been formed. At the tip of the granular magnetite cobs, the water flow must have been delaminated, and developed into a prolonged vortex. Underside this string shaped vortex the rust cob must keep growing (Fig. (**5-33**)).

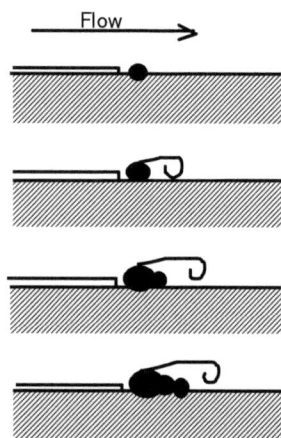

Fig. (5-33). A model illustrating the formation of vortex string at the top of nodular oxide followed by the formation of groove corrosion downstream [2].

2.6. Integrated Mechanism Description for Erosion-Corrosion in Carbon Steel and Low Alloy Steel

The important clue, which equaled the groove corrosion, to the mechanism was the specimen surface with the metallic luster which tended to appear after the tests in the water at higher temperatures and higher pH levels. This specimen surface must have been covered with a very thin but highly protective film of corrosion products which consisted of mainly hematite, Fe_2O_3, in almost the same composition as that of passivity film of stainless steel.

The fundamental factor for the generation of this film on steel surface is temperature: even on the surface of carbon steel, this film is formed when it reaches a certain critical temperature. Under this critical temperature, the magnetite, Fe_3O_4, deposits on the steel surface which is rather balky with lower mechanical strength, and similar, in this aspect, to the corrosion products on copper alloys. Therefore, when magnetite was deposited on the surface in the water of lower temperatures, the ring shaped damage occurred on the carbon steel specimen surface which was similar to that on the copper alloy specimen.

Next, this critical temperature receives the effects of various influencing factors. To begin with the dissolved oxygen content in the water (DO); the oxygen supply to the specimen surface tends to be hindered by the fixed vortex as it has already been pointed out in the previous chapter as to copper alloys. In addition to this, in the case of steels, the supply of the dissolved oxygen to the specimen surface was hindered by the string-shaped vortexes, which made it difficult for hematite film to form but enhanced magnetite to deposit along the vortex strings, thus the groove corrosion was formed. The next important influencing factor is pH level of boiler water: the higher the pH level, the lower is the critical temperature, resulting in the formation of hematite film on the steel surface, which was clearly demonstrated in Fig. (5-21). The influencing factor taken up finally is the alloying element: the inclusion of 1% chromium into the steel lowered the critical temperature as it can be recognized when Fig. (5-27) for the low alloy is compared with Fig. (5-22) for the carbon steel.

According to the description above, the dramatic changing behavior of the maximum depth of corrosion on carbon steel with an increasing pH (Fig. (5-23)) can be interpreted as follows:

1.   At pH 9. 0, the critical temperature was higher than 120 °C and consequently magnetite was deposited on the specimen surface which was damaged in a similar manner with copper base alloy, that is, ring-shaped damage due to shear as well as turbulence force on the ordinary flow specimen and a deep pit under the direct underside of the nozzle mouth on the reverse flow specimen;

2.   At pH 9. 5, the critical temperature was lowered for the specimen surface where sufficient oxygen supply was achieved, but it was still higher than 120 °C for those surfaces under the string-shaped vortexes where the supply of oxygen was so poor that hematite film covered the oxygen rich surface, but magnetite was deposited on the surface under the vortex resulting the formation of a macro-cell of corrosion with the macro cathode of former and the macro anode of latter. Since the surface area ratio was great, the corrosion proceeded in an anomalous rate;

3.   At pH 10. 0, the critical temperature was further lowered for all over the specimen surface irrespective vortexes, and the entire surface of specimen was covered with the film of hematite which resulted in the metallic luster just as that of stainless steel.

3. RELEASE FROM THE DILEMMA OF TESTING EQUIPMENT

It is an important intention of a testing methodology to obtain the test result in the shortest time with sufficient accuracy. In order to achieve this purpose, however, the test must be carried out under the condition which is different from that of the field. Then, a problem arises how close the testing methodology simulates the corrosion mechanism of components which consists of equipments and machines in the field. In order to examine whether the corrosion caused in the testing equipment sufficiently simulates the corrosion of the field, the ranking order of merit for various materials obtained by the test in the laboratory may be compared to that in the field. This was successfully demonstrated for the case of erosion-corrosion on copper alloys in the first section of this chapter. Most of the success was depended on the existence of materials ranking in the field; otherwise it would not have achieved the purpose.

In the second section of this chapter, the jet-in-slit tests were carried out on test specimens under various test conditions, namely various temperatures and pH levels of water, different velocities and direction of water flow *etc.* Since not all of those conditions imitate those of boilers in the field, these were rather acceleration tests in nature. Nevertheless, any damage under the condition of the field can be easily predicted, because the corrosion mechanism was obtained based on those test results which were appropriately interpreted by the engineers.

In conclusion, the point in utilizing an accelerated testing methodology is first to realize thoroughly the test conditions. Then, with the accelerated test results on the material, in addition to the comparison between the test condition and that of the field, the damage to material in the field can be accurately estimated. Thus, success totally depends on the engineer's ability.

REFERENCES

[1]   Matsumura M, Noishiki K, Sakamoto A. Jet-in-slit Test for Reproducing Flow-Induced Localized Corrosion on Copper Alloys. Corrosion 1998; 54: 79-88.

[2]   Tachibana S, Yabuki A, Matsumura M, Marugame K. Corrosion of Carbon Steel in Flowing Pure Water under High Temperature and High Pressure Conditions. Zairyo-to-Kankyo 2000; 49: 431-436.

[3]   Misawa T. The thermodynamic consideration for Fe-H_2O system at 25°C. Corros. Sci. 1973; 13: 659-676.

[4]   Ohguni T, Yabuki A, Matsumura M, Marugame K. Is increasing the pH of AVT Boiler Water Useful in Preventing the Corrosion of Carbon Steel? Zairyo-to-Kankyo 2001; 50: 386-389.

[5]   Yabuki A, Momikura H, Matsumura M, Marugame K. Corrosion of Low Alloyed Steel in Flowing Pure Water under High Temperature and High Pressure Conditions. Zairyo-to-Kankyo 2003; 52: 53-57.

CHAPTER 6

Case Study of Erosion-Corrosion in the Field

Abstract: Erosion-corrosion cases on various metallic materials were analyzed. Firstly, the erosion-corrosion damage in the pipeline system of pure copper was taken up, which was set up in the laboratory with the actual tubes and fittings obtained from the industrial market. Thus, the pure copper pipeline in the field was entirely reproduced except for the working liquid: a 1% copper chloride aqueous solution was used as the working liquid which had been proven in Chapter 5 to exert no influence on the corrosion mechanism, but successfully accelerated the corrosion rate. As a result, it was revealed that the erosion-corrosion occurred in the pipeline was not due to the separation of protective film through the shear force or the turbulence, but that it was a differential flow-velocity corrosion. Secondly, the case of differential flow-velocity corrosion in the pump casing of grey cast iron was examined. The cause of macro-cell formation or the localized graphitization corrosion on the pump casing near the shaft-hole was attributed to the characteristic transition of the graphitization corrosion process with the time. Thirdly, three similar cases of wall thinning in carbon steel pipe carrying pure water at elevated temperatures were taken up. The severe wall thinning in the pipe located at the downstream from the orifice or nozzle flow-meter was attributed indirectly to the shift of the electrochemical state of carbon steel surface from active to passive with the rise in the temperature, that is, the active/passive macro-cell corrosion discussed in Chapter 5. Two different types of wall thinning were found: uniform type and localized type. The former was brought about by the active/passive macro-cell due to the difference in oxygen supply, the latter by the similar macro-cell but due to the difference in the wall temperature. Lastly, the cause of the case which appeared to be an accidental or arbitral occurrence of erosion-corrosion in carbon steel pipeline was estimated to be some abnormal operation.

Keywords: Erosion-corrosion, pure copper, elbow, seawater pump, differential flow-velocity corrosion, gray cast iron, graphitization corrosion, carbon steel, boiler feed water, active/passive type macro-cell.

1. EROSION-CORROSION IN PURE COPPER PIPELINE

1.1. Introduction

Pure copper is used satisfactorily as a material of piping systems which carry various process liquids: the piping for the domestic supply of cold and hot water, the piping of the refrigerant liquid, heat transfer pipe tubes of various heat exchangers, the piping of ice making machinery, *etc.* This is not only due to its excellent heat conductivity but also due to the superior corrosion resistance which is given through the spontaneous formation of protective surface film of corrosion products under the service environments. Nevertheless, leaks have been experienced from time to time in the pipes at service. Severe damage was usually encountered at the bended parts (elbows and T joints) and the straight pipes located near the parts, which were simply classified as erosion-corrosion, and the cause was generally believed to be induced by shear force or turbulence of fluids with higher flow velocities. Accordingly, the proposed countermeasure for this sort of damage used to be reducing the flow velocity of the working liquid. However, it has to be taken into consideration that the erosion-corrosion damage on the copper-based alloys is brought about not always by the shear force or turbulence force, as it was thoroughly described in Chapter 4.

Murakami, the excellent postgraduate student, conducted a laboratory work of accelerated corrosion test on pure copper pipeline allowing the aqueous solution of copper chloride ($CuCl_2$) to flow through it. This testing condition might appear to be deviated from the actual operating conditions of various piping systems in the field. Nevertheless, the discussion on the test result has a significance equal to the case study of actual damage in the field, because the test apparatus was similar to the actual pipeline in the field and the test specimens were pipes and fittings obtained from the industrial market, and also because the corrosion mechanism was not to be affected by the test liquid as it had been proven in Chapter 5 [1].

1.2. Test Apparatus and Procedure

Fig. (**6-1**) schematically depicts the set-up of corrosion test loop used in the work. It consisted of a corrosion resistant tank, pump, flow meter, and the test section. The test solution was air saturated and set to 40 °C in the tank and then pumped into the test section through the flow meter. Thereafter it was returned back to the tank.

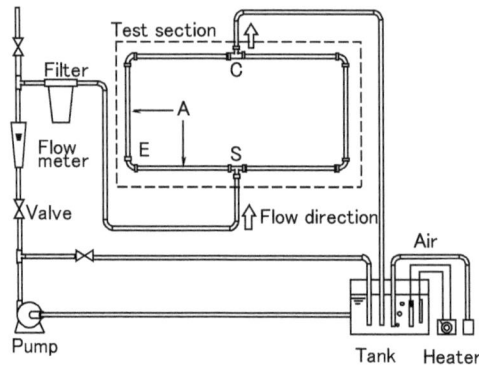

Fig. (6-1). Pure copper pipeline system used for measurement of wall thinning rate [1].

The commercial pure copper pipe tubes and fittings of standard product (JIS C1220; nominal diameter, 15A; standard outside diameter, 15. 88 mm) were used as the test specimens. The test section was composed of six straight pipes (500 mm in length), four elbows and two T joints. At the entrance as well as the exit of the test section, a 100 mm-long straight pipe was connected for the rectification of fluid flow. All the test specimens (pipes and fittings) were electrically conducted with each other, and the neighboring specimens were glued with epoxy resin, which easily released after the test run at the measurement of wall thickness. In the usual test runs, the flow velocity at the entrance of T joint was set to be 2 m sec^{-1} so that the flow velocity in the straight pipes and the elbows was to be 1 m sec^{-1}. The test liquid, a 1% $CuCl_2$ aqueous solution at pH 3. 5, accelerated corrosion rate 200-250 times higher than that in a 3% NaCl solution so that the test duration was no longer than 1 h.

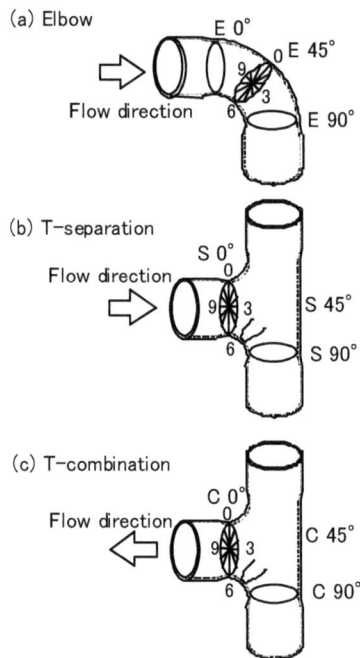

Fig. (6-2). Monitoring spots for wall thickness measurement: (a) Elbow; (b) T-separation; (c) T-combination [1].

Before and after the test run, the pipe wall thickness was measured using a dial gauge. In Fig. (**6-1**), the pipes and fittings are shown with arrows where the wall thickness was monitored: A, straight pipes; E, an elbow; S and C, T joints. Fig. (**6-2**) shows the monitoring spots around the fittings monitored: for the elbow, the wall thickness monitoring was conducted at three spots: E 0°; E 45°; and E 90°, and 12 measuring points were chosen at each spot with 15° intervals over inside wall surface (Fig. (**6-2(a)**)). Fig. (**6-2(b)** and (**c**)) depict the monitoring spots on each T joint. Along the straight pipe, twelve monitoring spots were located at 50 mm interval (25, 75, · · · 425 and 475 mm) in both side upstream and downstream from that elbow which is indicated as E in Fig. (**6-1**).

1.3. Test Results

Preliminary tests were carried out on the straight pipes. Firstly, the relationship between the mass loss rate (mg mm^{-2} h^{-1}) of straight pipe and the flow velocity of test solution was examined (Fig. (**6-3**)). The mass loss rate monotonously rose with flow velocity up to 2. 6 m sec^{-1} without any sudden jumping up of the rate. It was, therefore, able to be estimated in considerable probability that it must not be the case of erosion-corrosion in which the corrosion damage originated from the separation of the protective surface film, and measurements thereafter might be conducted at any flow velocity under 2. 6 m sec^{-1} without taking care of the breakaway velocity.

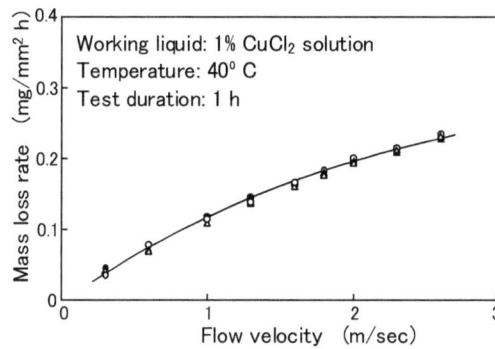

Fig. (6-3). Relationship between mass loss rate and flow velocity for straight pipe.

Secondly, the circumferential distribution of wall thinning rate was examined in the straight pipes at 25 mm downstream as well as upstream from the elbow (Fig. (**6-4**)). It can be seen in the diagram that the wall thinning rate was constant over the circumferential points at the spot 25 mm upstream from the elbow. On the other hand, at the spot 25 mm downstream from the elbow, it varied over the circumferential points yielding the maximum rate at the measuring point 6, that is, interior side of the elbow, and the minimum rate at several points in the exterior side of the elbow at which the level was comparable to that constant rate at the spot 25 mm upstream from the elbow.

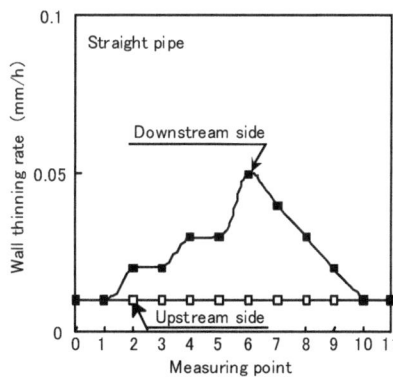

Fig. (6-4). Distribution of wall thinning rate in cross sections of straight pipe: 25 mm downstream and upstream from the elbow [1].

Fig. (**6-5**) summarizes the observed wall thinning rate distribution at measuring point 6 along the axis of the straight pipes set upstream and downstream from an elbow as well as in the elbow in between. It should be noted that the wall thinning rate was constantly maintained in the straight pipe set upstream, but rapidly rose at the entrance of the elbow, and then fell down as it approached the exit of it, showing further decay in the straight pipe as the distance from the elbow exit increased. Thus, the constant level of wall thinning rate measured in the straight pipe located at the 25 mm upstream from the elbow was taken as the reference level of the wall thinning rate.

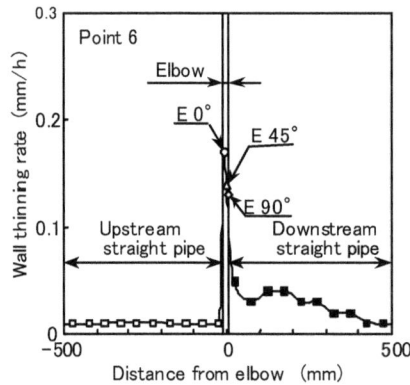

Fig. (6-5). Distribution of wall thinning rate along straight pipes located upstream and downstream from the elbow as well as in the elbow in between [1].

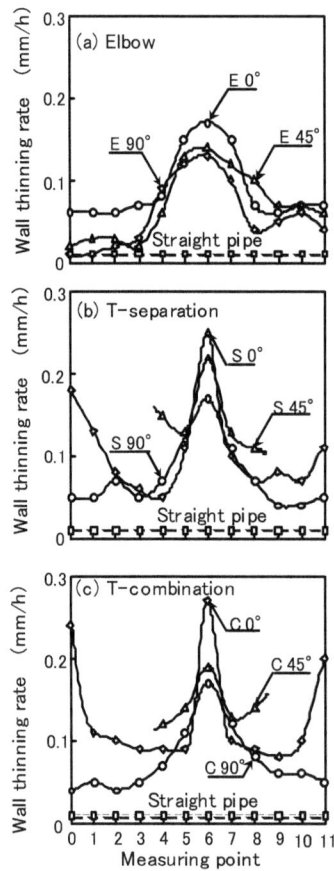

Fig. (6-6). Distributions of wall thinning rate in fittings: (a) Elbow; (b) T-separation; (c) T-combination [1].

Fig. (**6-6**) compares the circumferential distributions of wall thinning rate measured at the fittings: (a) Elbow; (b) T-separation; (c) T-combination. Together with these distribution profiles in fittings, the wall thinning rate measured in the straight pipe set upstream from the elbow is also shown for the reference. It is evident in Fig. (**6-6(a)**) that the rate of wall thinning at any measuring point of the elbow was higher than the reference level of the straight pipe. Within the elbow, the highest wall thinning rate was observed at measuring point 6 (elbow interior side) at every spot: E 0°, E 45°, and E 90°. The wall thinning rate at point 6 in the elbow was especially high at E 0° yielding 0. 17 mm h^{-1}. Similarly, the highest wall thinning rate at every monitoring spot in T joint, either separating or combining, was observed at point 6. At the monitoring spot S 0° (Fig. (**6-6(b)**)) and C 0° (Fig. (**6-6(c)**)), the wall thinning rate reached the maximum not only at the measuring point 6 but also at point 0. The measuring point 0 in the spot S 0° as well as in the spot C 0° was actually equivalent to the point 6, that is, the interior side wall of the elbow.

And thus it seemed allowable to claim that the highest wall thinning at any given monitoring spot of the elbow and the T joint would occur always on the interior side wall of the fittings. The maximum wall thinning rate at point 6 was higher for the T joints than for the elbow. It can be understood that this was due to the two times higher average flow velocity at S 0° and C 0° in the respective T joint than that at elbow entrance, E 0°.

1.4. Erosion-Corrosion Mechanism at Pipe Bends

The flow patterns in the pipe bends have been intensively investigated in the field of the fluid dynamics. Fig. (**6-7**) reproduces the representative results of such study made for an elbow cross section.

Centrifugal force is yielded for the fluid flow in the elbow, and accordingly the fluid would be pushed towards exterior side of the elbow surface. Consequently, fluid of the exterior side would flow down to the interior side of the elbow yielding circulating secondary fluid flow (lower semicircle in Fig. (**6-7**)). This would lead to the shift of axial stream line towards the elbow exterior side to yield the emergence of the fast flow zone near the exterior wall of elbow (upper semicircle in Fig. (**6-7**)). The flow velocity is actually zero at any point of the pipe wall surface, but, when the comparison is made at a comparable distance from the pipe wall surface, the flow velocity would be lower on the interior side surface (measuring point 6) than on the exterior side surface. In the experimental work described above, deeper wall thinning was detected for the interior side surface. That is, the severer corrosion occurred at the lower flow velocity side, and accordingly the detected corrosion to the elbow was identified to be differential flow-velocity corrosion.

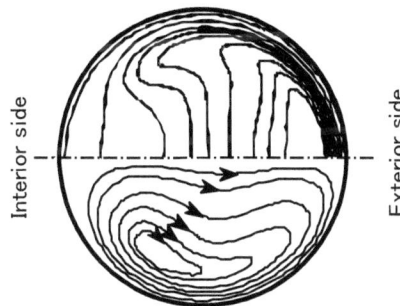

Fig. (6-7). Fluid flow in the cross section of bend: equal-velocity lines (upper half) and secondary flow stream lines (lower half) [1].

1.5. Protection against Differential Flow- Velocity Corrosion at Pipe Bends

Slowing down of the flow velocity is usually recommended as the countermeasure against the erosion-corrosion in elbow. However, such measures cannot be introduced in this case, because the wall thinning cannot be attributed to the shear force or turbulence of the fluid which flows at high velocity, but to the difference in the flow velocity. The equalization of flow velocity, alternatively, would be sufficient to reduce the corrosion damage. According to Fig. (**6-7**), the flow velocity difference in the elbow appears to emerge as the consequence of centrifugal push of the fluid against the exterior side surface of the elbow and the resultant loss of the flow near the interior side surface. Thus, in order to diminish the flow velocity difference in the elbow, a guiding vane of styrol resin 0. 15 mm in thickness, as depicted schematically in Fig. (**6-8**), was inserted in the elbow. The test results obtained at the spot E 0° with this guide vane are presented in Fig. (**6-9**) together with the result without the vane. The measurements in the diagram showed that, by introducing the guide vane, the outstandingly severe wall thinning on the interior side surface of the elbow was diminished, being replaced by the smaller wall thinning in cyclic pattern with smaller period over the circumference of monitoring spot.

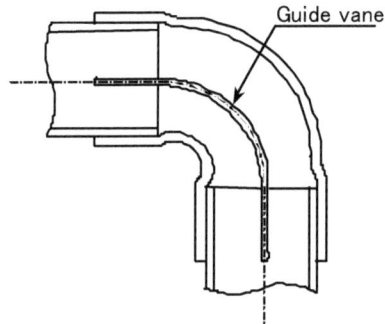

Fig. (6-8). Guide vane installed for suppressing the distribution of flow velocity in elbow [1].

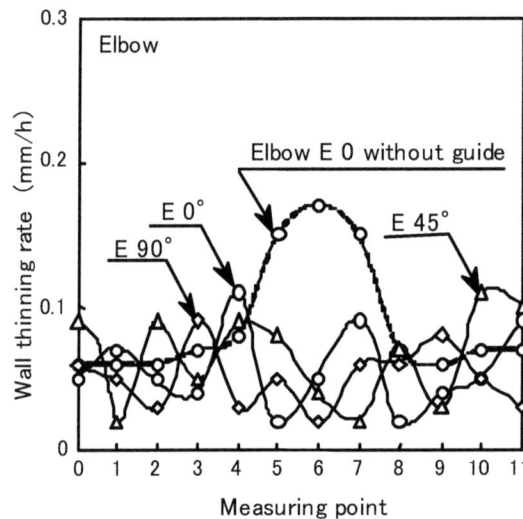

Fig. (6-9). Distribution of wall thinning rate in the elbow installed with a flow guide vane [1].

2. DIFFERENTIAL FLOW-VELOCITY CORROSION ON SEAWATER PUMP CASING OF GREY CAST IRON

2.1. Background

As it was described in Chapter 1, Kitajima *et al.* demonstrated, with a unique technique, that localized corrosion originated near the shaft hole of seawater pump casing was macro-cell corrosion in nature, and gave the terminology of "differential flow-velocity corrosion" for the first time to the localized corrosion of this mode (Fig. (**6-10**)).

However, they have not yet clarified the formation mechanism, or the reason why the macro-cell originated in the pump casing of grey cast iron. In the case of copper base alloys, the formation process of this sort of macro-cell, namely the generation mechanism of differential flow-velocity corrosion, has already been elucidated in Chapter 4 by using the jet-in-slit to find out that more than a single process existed for the formation of the macro-cell. Furthermore, the formation processes greatly depended on the characteristics of the copper base alloy and the flow conditions as well, so that it cannot be expected that some of the macro-cell formation mechanism for copper alloys be applied to the pump casing of grey cast iron. Here in this section of this chapter, it is taken up as the object of case study how the macro-cell of corrosion was formed in the seawater pump casing of grey cast iron.

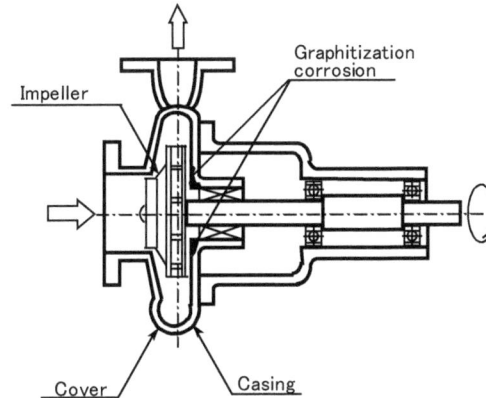

Fig. (6-10). Graphitization corrosion occurred near the shaft hole of seawater pump casing made of grey cast iron.

2.2. Micro-structure of Grey Cast Iron and Graphitization Corrosion

The micro-structure of grey cast iron consists of the matrix of pearlite and scattered graphite in it, and the pearlite the lamination of ferrite and cementite (Fig. (6-11)). When it corrodes, the ferrite dissolves out as a ferrous ion, that is Fe^{2+}, but the graphite as well as the cementite remains in the metal as it is. As a consequence, the residue layer or graphitization layer, which is more abounding in carbon component than the base metal, is originated in the surface of the cast iron. Hence, the corrosion of grey cast iron is generally called as graphitization corrosion or skeleton corrosion. The strength of this residue layer is almost zero.

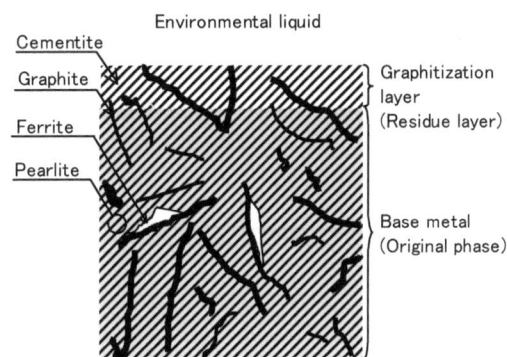

Fig. (6-11). Microstructure of grey cast iron and graphitization corrosion (schematic illustration).

2.3. Transition of Corrosion Process

In Fig. (6-12), the development of the residue layer in the surface of block of grey cast iron is schematically illustrated (upper), together with the time dependence of Fe^{2+} ion dissolution rate, that is, the corrosion rate (lower). According to Suezawa and Shinohara [2], the corrosion of grey cast iron passes through some processes as depicted in the illustration: in the initial stage of corrosion, the graphitization layer originates in the cast iron surface and rapidly extends its range over the surface. With this, the dissolution rate of ferrous ion rises rapidly (A→B), since the graphitization layer functions as an active cathode. At the time point B when the graphitization layer covers the whole surface, the increase in the ferrous ion dissolution rate is terminated, and then it turns to decrease with the time thereafter, since the elution of the ferrous ion is inhibited with the increase in thickness of the layer (B→C). As the corrosion rate in the first process is higher, the transition of process occurs earlier, and the corrosion rate in second process sinks sooner.

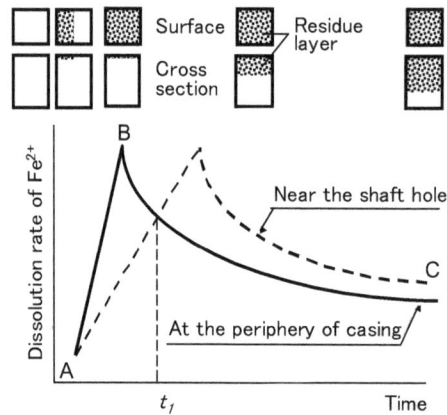

Fig. (6-12). Development of residue layer in the surface of grey cast iron and transition of corrosion process [2].

2.4. Formation of Macro-Cell of Corrosion

According to the transition of corrosion process of grey cast iron illustrated in Fig. (6-12) the occurrence of the localized graphitization corrosion on the seawater pump casing near the shaft hole may be estimated as follows. In the peripheral surface of the pump casing, the fluid flow velocity was higher, and accordingly the corrosion rate was higher, so that the transition in the corrosion process occurred earlier as indicated with the solid line in the diagram. In comparison with this, the fluid flow velocity was lower near the shaft hole, and accordingly the corrosion rate was lower, and the transition was retarded as shown with the broken line in the same diagram.

This retardation in the transition brought about the situation that at the time point, t_1, the corrosion rate near the shaft hole reached at that of peripheral part, and their relative position was reversed afterwards. The difference in the corrosion rate between those parts in the casing was followed by the formation of such macro-cell that consisted of the macro-anode near the shaft hole and the macro-cathode in the peripheral part of the pump casing. The macro-cell current flowed from the peripheral part into the restricted area near the shaft hole, just as it was detected by Kitajima *et al.*, resulting in the severe localized damage of graphitization corrosion at the place. Here, it should be noted that the time duration from the beginning of corrosion until t_1 in the diagram was intentionally prolonged in order to illustrate the process in detail.

3. WALL THINNING OF CARBON STEEL PIPES IN BOILER

3.1. Overview of Boiler Pipeline Incidents

The pipe specifications, environmental and operating conditions of the pipelines which have been taken up here as the examples of the case encountered incident during the last twenty years in Japan are listed in Table **6-1**.

In the first case, in 1993, a rupture occurred due to the wall thinning in carbon steel pipe which had been transporting pure, high-temperature water in the waste heat boiler plant since 1977. It was located at downstream from an orifice flow meter installed in line B, one of the two pipelines A and B, which had been in parallel operation under the same operating conditions. The sketch of the damage in the pipeline B and the residual wall thickness distribution along the pipe is shown in Fig. (6-13). The wall thickness decreased gradually from measuring points 5 to 0, to reach 0. 63 mm (the minimal among the measurements) at the end of the pipe where it was welded to the flange. On the other hand, the distribution of the residual thickness in the circumferential direction at each measuring point was almost uniform, as shown in the same table. The maximum average wall-thinning rate of 0.54 mm y^{-1} was obtained from the minimal residual thickness cited above and the operating duration of 15 years. In pipeline A, which was parallel with pipeline B, the maximum wall thinning was 5. 4 mm so that the maximum wall thinning rate for this pipeline was 0.36 mm y^{-1}.

Table 6-1. Operation conditions

	Case 1 (TA)	Case 2 (KO)	Case 3 (MI)
Specification of pipe	STPT370	STPT38S	SB42
Outside diameter (mm)	114.3	165.2	ca.560
Inside diameter (mm)	97.1	136.6	538.2
Original wall thickness (mm)	nom. 8.6	nom.14.3	nom.10
Temperature of liquid (°C)	123	145	142
Pressure of liquid (MPa)	-	11.2	0.93
Flow rate (t/h)	-	76	ca.1700
Flow velocity (m/s)	1.9	1.6	ca 2.2
pH	8.5-9.5	9.2	8.6-9.3
Dissolved oxygen conc. (ppb)	< 5	-	< 5
Operation duration (year)	15	18	21
Wall thinning rate (mm/y)	0.36 / 0.54	0.78	0.45 / 0.33

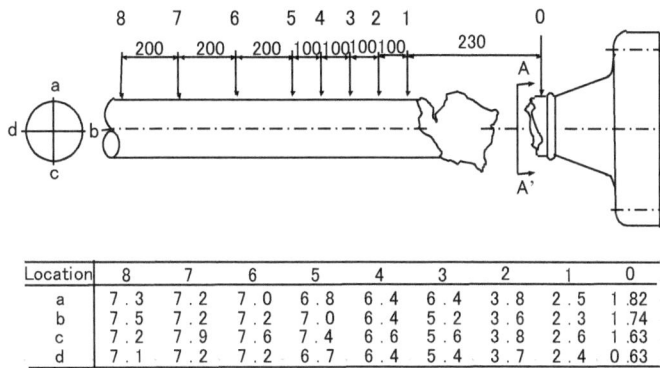

Location	8	7	6	5	4	3	2	1	0
a	7.3	7.2	7.0	6.8	6.4	6.4	3.8	2.5	1.82
b	7.5	7.2	7.2	7.0	6.4	5.2	3.6	2.3	1.74
c	7.2	7.9	7.6	7.4	6.6	5.6	3.8	2.6	1.63
d	7.1	7.2	7.2	6.7	6.4	5.4	3.7	2.4	0.63

Fig. (6-13). Ruptured pipe in case 1 and residual wall thickness distribution: hot water flow from right to left.

The second case of incident arose in the utility boiler of a petroleum refining plant. Wall thinning caused carbon steel piping to rupture downstream from a flow nozzle that had been transporting pure, high-temperature water. All the conditions under which this piping was operated are also listed in Table **6-1**. It should be noted that the environmental temperature of 145 °C falls into the range within which the wall thinning has frequently occurred in carbon steel pipe. Although the concentration of dissolved oxygen (DO) is not described, the environmental liquid was boiler feed water under the all volatile treatment (AVT). Assuming the wall thinning depth was equal to the nominal pipe wall thickness, the thinning rate was determined to be 0. 78 mm y^{-1}.

The cross-sectional view and the top view of the ruptured pipe are shown in Fig. (**6-14**). The square hole in the ceiling of the pipe in the top view resembles the square fragment cut out of the pipe wall of which sketch is attached nearby. In the cross-sectional view in the figure, this piece appears to be described being rotated by 90°. In this spot, the heat insulation layer over the external surface of the piping appears to have been cut off squarely to create a window for the inspection of pipe wall thickness. It is possible that there was some clearance between the circumference of the lid and the opening in the heat insulation layer over the piping. Heat leakage might have occurred through the clearance in the interval between the inspections. As a result, a square shaped cold belt may have arisen along the clearance, which might have had something to do with the wall thinning there.

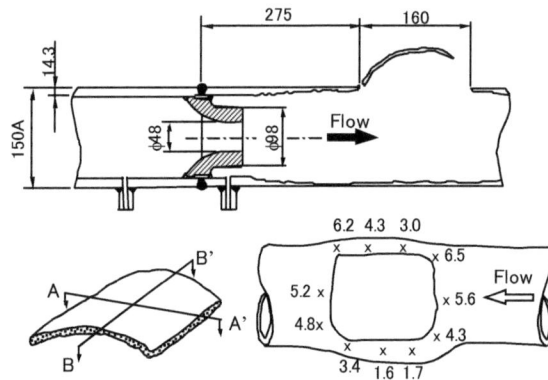

Fig. (6-14). Cross-section of the ruptured pipe of case 2: hot water flow from left to right.

The third case occurred in August, 2004 at Mihama nuclear power plant. The following figures and diagrams are quoted from the reports issued by the investigation committee of the Nuclear and Industrial Safety Agency Japan (NISAJ) [3].

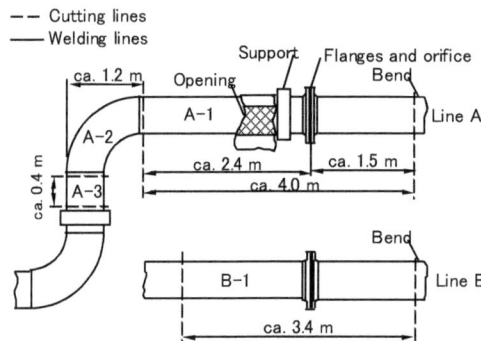

Fig. (6-15). Pipelines of case 3, hot water flow from right to left, cited from the documents of the Nuclear and Industrial Safety Agency Japan (NISAJ) [3].

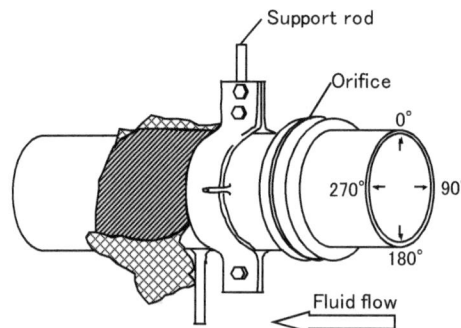

Fig. (6-16). Appearance of pipeline A of case 3, released from NISAJ [3].

Two lines of piping, A and B, as shown in Fig. (**6-15**), had been driven in completely identical conditions (Table **6-1**). Wall thinning occurred in the lower reach of the orifice meter of pipeline A, and it ruptured as shown in Fig. (**6-16**). Comparison between pipelines A and B in the circumferential distribution of residual wall thickness reveals that in pipeline A the wall thinning appears to be biased toward the ceiling of the pipe, while the wall thinning in pipeline B is almost circumferentially uniform except for the bottom part (Fig. (**6-17**)).

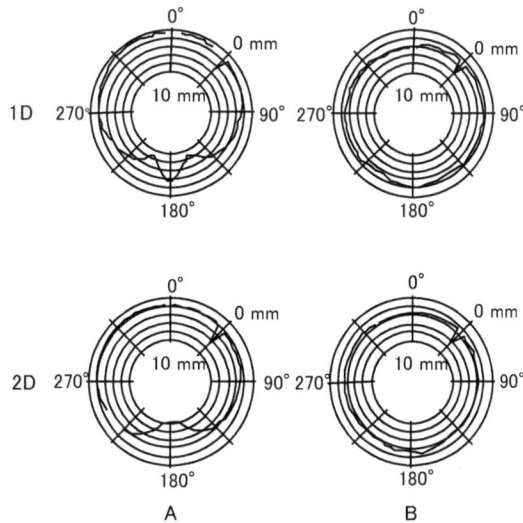

Fig. (6-17). Residual thickness distribution diagram for pipe walls at cross sections located downstream from orifice by the distance 1D and 2D (D, inside diameter of pipe), NISAJ [3].

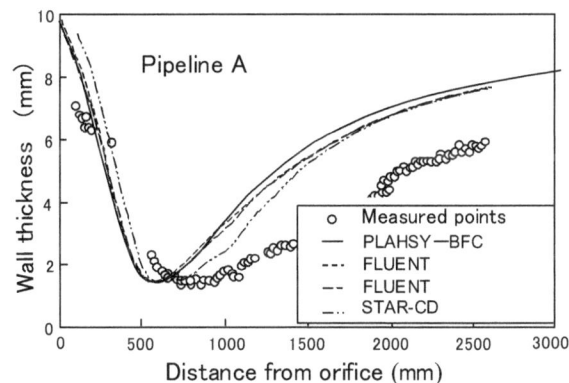

Fig. (6-18). Distribution of residual wall thickness along the pipe axis at 270° shown in Fig. **6-16**: points, measured values; lines, calculated values, NISAJ [3].

The fact that the bias occurred only in pipeline A, where a supporting rod was installed, led to the conjecture that it might have originated from the lowering of the pipe wall temperature by heat radiation from the support rod. The wall thinning depth distribution along the pipe axis of line A at the circumferential position of 270° is shown in Fig. (**6-18**), which was almost identical with that of pipeline B.

3.2. Categories of Wall Thinning

Some common features can be found in the damage described in the three cases above. First, all of the wall thinning occurred on the pipe wall downstream from the orifice or flow nozzle. The flow velocities upstream from the flow-meters were around 2 m sec^{-1}, which were not unusual high levels. The range of pH of the environmental liquids was 8. 5-9. 5 and DO was less than 5 ppb. All of the liquids were AVT water.

The water temperatures, except for the first case, remained in the range of 140-150 °C, in which wall thinning rates have peaked, as described in Chapter 1. There was a conjecture that in case 1, temperature measurements at other positions in the pipeline had been substituted for those in the area of wall thinning, and it was thought to be approximately 140 °C.

Second, the examples of wall thinning described above may be categorized into two types, as follows. One is uniform wall thinning generated over a comparatively wide range, which is observed in every pipeline of cases 1 through 3. Another type was generated in the localized area with a comparatively higher rate, which became the cause of pipe rupture. This was observed in case 2 (Fig. (**6-14**)) and in pipeline A of case 3 (Fig. (**6-15**)). Hereafter, the former will be called "uniform wall thinning" and the latter "localized wall thinning ". These types have other features in addition to these mentioned above.

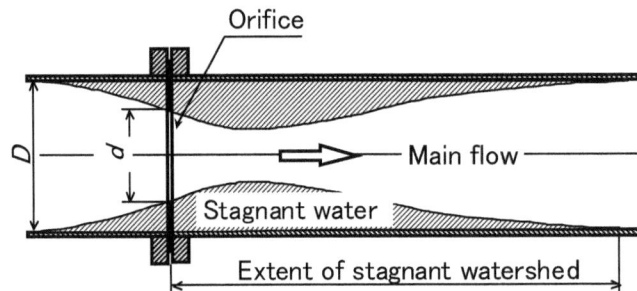

Fig. (6-19). Stagnant watershed originated downstream from orifice; *l*, extent of stagnant watershed [4].

As to the uniform wall thinning type, a close relation was recognized between the extent of the thinning area and the inside diameter of pipe. According to the fluid dynamics, as shown in Fig. (**6-19**), the extent of stagnant watershed originated downstream from a nozzle or orifice meter is a function of the opening ratio of the orifice, *m*, where $m = (d/D)^2$; *d*, diameter of orifice mouth; *D*, inside diameter of pipe: the smaller *m* is, the lager the extent is, and it approaches 6 times the inside diameter of pipe as *m* decreases [4]. As listed in Table **6-2**, the ratio of the extent of uniform wall thinning area, *l*, for the inside diameter, *D*, is almost 6. That is to say, the area of uniform wall thinning almost agrees with the stagnant watershed which arises downstream from the orifice.

Table 6-2. Extent of uniform wall thinning

	D (mm)	*m* (-)	*l* (mm)	*l/D* (-)
Case 1 (TA)	97. 1	0. 48	630	6. 4
Case 2 (KO)	137	0. 12	735	5. 4
Case 3 (MI)	520	0. 39	3000	5. 8

On the other hand, a feature of localized wall thinning is its higher rate of thinning as compared with uniform wall thinning, when they occur under equal conditions. Another important feature is that it has occurred in those places where the temperature was seemingly lower than the surroundings.

3.3. Passivity of Carbon Steel at Elevated Temperatures

Passivity is the property underlying the corrosion resistance of many useful metals and alloys, such as aluminum and stainless steel. Some metals and alloys can be made passivity by the exposure to passivity environments, for example, iron in chromate or nitrate solutions. This occurs because a very stable passive film (mainly consisting of hematite, Fe_2O_3) forms on the surface.

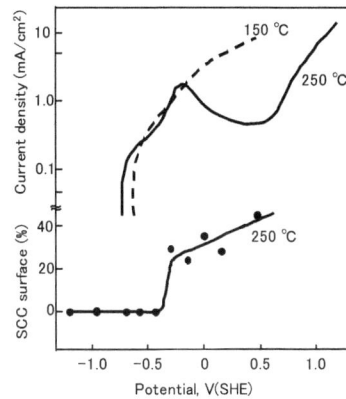

Fig. (6-20). Anodic polarization curves for carbon steel at 150 and 250 °C in a solution of 0. 1 M-Na$_2$SO$_4$ with a DO of 200 ppb [5].

Using an autoclave, Umemura and Kawamoto [5] conducted a laboratory test on the effect of environmental liquid temperature on the passivity of carbon steel in a 0. 1M-Na$_2$SO$_4$ aqueous solution with dissolved oxygen of 200 ppb. The results of the experiment are shown in Fig. (**6-20**): the carbon steel appeared active at 150 °C, and in passivity at 250 °C. An accurate observation of the electrochemical polarization curve found the gradient of the curve at 150 °C had risen somewhat higher than that at lower temperature; thus, the passivity of the carbon steel at high temperatures seems to gradually advance from per 150 °C. The metal must be in the situation where it is not completely but by half in passivity, which is usually called as passivation.

The passivity of carbon steel is widely believed to be dependent on the concentration of the oxidizing agents and pH of the environmental liquid, but is not influenced by the dissolved oxygen concentration (DO). However, this view is taken from the cases where the dissolved oxygen concentration has been sufficiently high. Recently, Bouvier *et al.* [6] conducted experiments in the CIROCO loop of EDF (Electricité de France) that allowed the test liquids (180 and 235 °C) to run through a carbon steel tube with 8 mm inside diameter at flow velocities of 5 and 10 m sec^{-1}. As a result, it was confirmed that, when it is in pure water, carbon steel must have a DO of more than 1 ppb at its surface so that it can be protected against FAC. These researchers hypothesized that this was due to the formation of hematite in the void of the magnetite layer on the test specimen surface, because the potential of the specimen rose when the DO concentration of the environment exceeded 1 ppb. Thus, their experimental results seem to indicate that at least 1 ppb of dissolved oxygen is necessary for the passivity of carbon steel even in high temperature.

3.4. Generation Mechanism of Uniform Wall Thinning

Combining various experimental results described above and the integrated mechanism description for erosion-corrosion in carbon steel which was given in Section 2 of Chapter 5, the effect of temperature on the corrosion rate of carbon steel in pure water can be depicted as shown in Fig. (**6-21**).

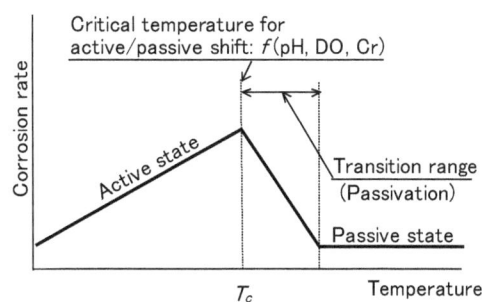

Fig. (6-21). Temperature dependency of corrosion rate of carbon steel in the range of active/passive shift: T_c. shift-onset critical temperature.

Because a corrosion process is fundamentally a chemical reaction, the corrosion rate of carbon steel in pure water rises with temperature. When it reaches a certain temperature it begins to become passive, which is passivation as described above, and at higher temperature becomes completely stabilized, that is, passivity. The characteristic temperature at which the shift from active to passive begins may hereafter be referred to critical temperature for active/passive shift and is symbolized as T_c. This characteristic temperature may depend on environmental conditions, or may be a function of the pH and the DO of the liquid, and of the alloy element content in the steel. It appears to be 140-150 °C for plain carbon steel pipes in pure AVT water.

Based on the schema in Fig. (**6-21**) the generation mechanism of uniform wall thinning in the incidents cited in Table **6-1** can be explained as follows. The temperature of liquid, T_e, and accordingly the temperature of the pipe wall were much higher than the critical temperature of active/passive shift, T_c, so that the inside wall surface of the pipe was in stabilized passivity, as indicated by the solid square in Fig. (**6-22**). In the stagnant watershed downstream from the orifice, however, the supply of dissolved oxygen to the pipe wall surface was insufficient, and the oxygen concentration was lower than that in other surfaces. The critical temperature of the wall surface under the watershed was consequently raised higher, that is T_c' in the figure, and the corrosion rate was rather high because it was not in complete passivity but in passivation (solid circle in Fig. (**6-22**)). Thus, two surfaces with different corrosion rates came together, which resulted in the formation of a macro-cell with the macro-anode of the wall surface under the stagnant water and the macro-cathode of the surrounding surface, following the protocol described in detail in Chapter 4. And, the corrosion rate in the watershed was accelerated, as indicated by the open circle in the figure.

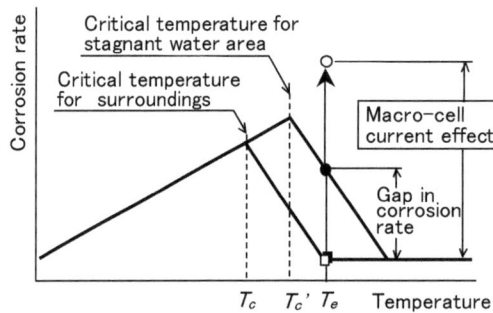

Fig. (6-22). Gap in the critical temperature between stagnant watershed area and its surrounding, which leads to the generation of another gap in corrosion rate, followed by the formation of macro-cell.

3.5. Generation Mechanism of Localized Wall Thinning

The generation process of the localized wall thinning that appeared in the middle of the uniform wall thinning field in cases 2 and 3, cited in Table **6-1**, may be explained by the schema in Fig. (**6-23**), where the pipe wall temperature at the location of issue, T_s, was added to the preceding schema in Fig. (**6-22**).

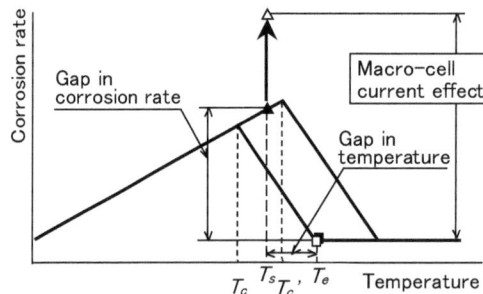

Fig. (6-23). Gap in temperature between the stagnant water area and the surrounding area, T_s vs. T_e, leading to the gap in corrosion rate followed by the formation of macro-cell.

Because of considerable heat leakage, the temperatures at the wall surface next to the wall-thickness-inspection window as well as the pipe supporting rod must be considerably low, compared to those of their surroundings, so that these locations were in active state rather than passive state (solid triangle symbol in Fig. (**6-23**)). This temperature gap across T_c' produced the gap in the corrosion rate between them, which resulted in the formation of a macro-cell that consisted of the macro-anode of the surface of the lower temperature and the macro-cathode of surrounding surfaces of the higher temperature. The wall thinning progressed at such a high rate, as indicated by the open triangle in Fig. (**6-23**), that it caused the pipes to finally rupture.

The potential *vs.* current density diagram in Fig. (**6-24**) illustrates, from an electrochemical viewpoint, the formation process of macro-cell for the case. Here, the reductive reaction of hydrogen ions is assumed to work as the cathode reaction. This is because the reductive reaction of oxygen is not feasible to occur under the stagnant watershed of AVT water, where the DO is extremely low. For the simplicity, the hydrogen ion reduction reaction is assumed not to be dependent on the flow conditions, so that it is deemed common to the three surfaces. On the other hand, three different states of anode reaction are assumed as described above: active and passive state as well as passivation in between.

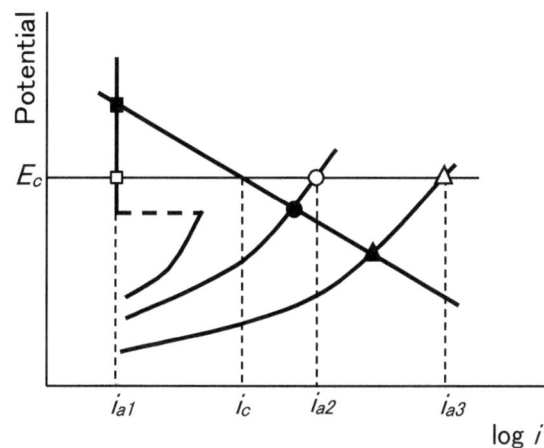

Fig. (6-24). Potential *vs.* current density diagram illustrating the formation of macro-cell with three electrodes: surroundings (passivity), stagnant water area (passivation), and locally cooled area (active state).

When the surfaces independently corrode, they take their own positions in the diagram at the points where the cathodic polarization line intersects the anodic lines, which are indicated by those three different symbols (solid square, circle and triangle), and their corrosion potentials and corrosion rates are respectively at their coordinate positions. In the actual pipe wall surface their potentials ought to be at a single and the same level because they are electrically short-circuited. At the same time, the relation of Eq. (6-1), namely, "the total of cathodic currents = the total of anodic currents", ought to be also established to obey the law of conservation.

$$3I_c = I_{a1} + I_{a2} + I_{a3} \tag{6-1}$$

Obeying to the regulations above, each potential must move to the level which is common to the surfaces, as indicated by E_c in the ordinate and the open symbols in the diagram. At the same time, the macro-cell currents must be exchanged among the macro-electrodes (those surfaces) in order to establish the balance of Eq. (6-1), which resulted in increases in corrosion rates, as indicated by i_{a2} and i_{a3}. This is the macro-cell current effect described in Chapter 4. In addition to this, the localized wall thinning rate might rise much higher than that of uniform wall thinning, because the surface area of localized cooling was smaller than that under the stagnant water, and far smaller than those areas of passivity. This is the surface area ratio effect described also in Chapter 4.

3.6. Accidental Occurrence of Erosion-Corrosion

One of the features of the wall thinning in the carbon steel pipeline due to erosion-corrosion is the accidental occurrence, as it was already pointed out in Chapter 1: it sometimes occurs and sometimes does not, even when the related parameters are identical. The typical example cited in Chapter 1 was the case below, where three boilers of the identical set-up were operated in parallel under the identical conditions.

Around the end in 1980's, the incident occurred in the pipeline which transported hot AVT water at 150 °C to a cooling unit attached to the boiler. In the normal operation, the flow velocity of the water varied in the cyclic mode as illustrated in Fig. (6-25): it was allowed to rise linearly with time from 0 up to 8 m sec^{-1} and then fall down to zero, and this cyclic process was continuously repeated afterwards. Within four years after the beginning of commercial operation, the pipeline was suddenly ruptured due to the serious decrease in pipe wall thickness (Fig. (6-26)). The wall thinning rate calculated from the nominal wall thickness of the pipe and the operation duration was as high as 1. 2 mm y^{-1}.

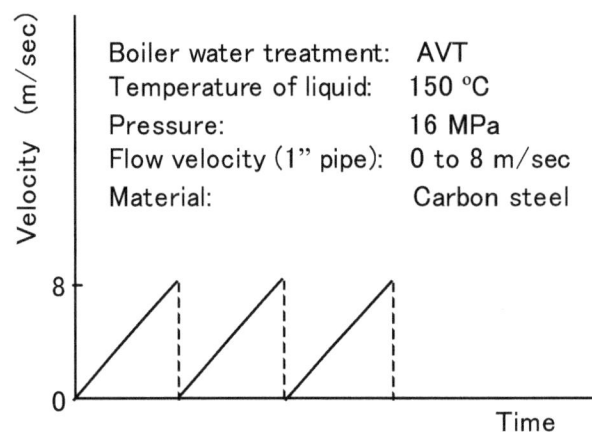

Fig. (6-25). Operation conditions and the cyclic variation in flow velocity.

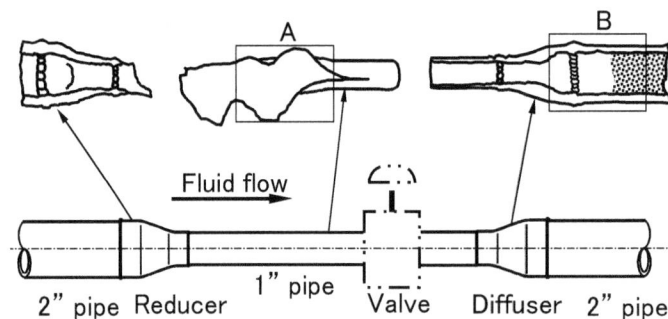

Fig. (6-26). Arrangement of pipes and fittings in the pipeline where the rupture occurred.

The inspection of the ruptured pipe and the investigation of background of incident such as service history of the boiler *etc.* could not reach any conclusion, but the following problems and questions remained unsolved: the flow velocity was not so high to separate the oxidation product film, hematite or magnetite it might be, from the metal surface since these films on the carbon steel are usually hard and rigid; the serious wall thinning occurred not downstream from the valve where turbulence used to develop in the fluid flow, but downstream from the reducer where the turbulence of fluid flow suppressed due to the acceleration of the flow velocity. Any stagnation watershed could not occur in the pipeline. The most serious problem was the reason why only one of the three boilers of the same set-up with the same operation history encountered this misfortune, other two boilers being completely free from it.

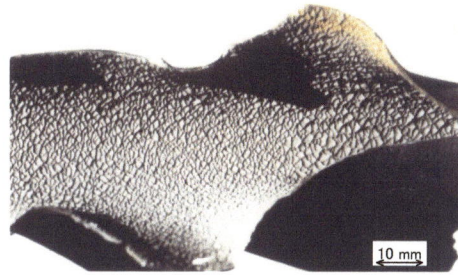

Fig. (6-27). Macrograph of the inside wall surface at location A in Fig. (**6-26**).

Fig. (6-28). Macrograph of the inside wall surface at location B in Fig. (**6-26**).

More than ten years later when the macro-cell corrosion mechanism began to be applied to the problems of erosion-corrosion, the clue to the solution of problems and questions mentioned above as well as the conclusion was found, which lay not in the corroded pipe wall surface (Fig. (**6-27**)), but in the sound wall surface without any thinning in thickness: the red collared thin film of deposit on the wall surface which was located considerably downstream from the ruptured pipe (Fig. (**6-28**)).

It must be γ-Fe_2O_3, that is, lepidocrocite which is a similar chemical compound to hematite, Fe_2O_3, and originates at higher temperature than magnetite, Fe_3O_4, does. In the same macrograph, magnetite can be recognized as the film of dark-brown deposit. The most important is the relative position of these deposits of chemical compound as to the water flow direction which was from left to right as shown in the figure: the magnetite deposit on the left side, that is, upstream from the deposit of lepidocrocite, which is the clear evidence for being at a lower temperature of the upstream side wall than of the downstream side wall. This can never happen as long as the fluid flows. Or, in other words, it is the decisive evidence of that the fluid was once stopped or stagnant, during which the upstream side pipe wall must be badly chilled, probably due to the insufficient heat insulation, and the temperature must be lower than the critical temperature so that the pipe wall surface became active. In the meanwhile, the temperature of the downstream side pipe wall could remain higher than the critical temperature, which made it in a more stable state resulting in the deposition of oxide with a similar chemical composition with hematite. The macro-cell of corrosion must have consisted in the same process that was illustrated in Fig. (**6-23**), and wall thinning progressed at a high rate at the macro-anode.

The reason why the macro-cell of corrosion occurred only in the concerned boiler but not in other two may be estimated as follows: once in the early stage of commercial operation, the water flow stopping period between the cyclic flow velocity processes (Fig. (**6-25**)) was accidentally and irregularly held longer for the pipeline of the concerned boiler.

3.7. Characteristic Appearance in Erosion-Corrosion Damaged Surface of Carbon Steel

Characteristic planes appeared in the inside wall surface of the carbon steel pipe (Fig. (**6-29**)), in particular in the place where deep wall thinning occurred, which ruptured in Mihama nuclear power plant, as described in the beginning of this section (Fig. (**6-16**)). The members of the investigation committee of the Nuclear and Industrial Safety Agency (NISA) gave the planes the name of "fish-scale pattern", and considered that this arose from the effect of flowing fluid.

Fig. (6-29). SEM micrograph of fish-scale pattern released by NISA [3].

On this characteristic appearance, the author gave the comment of the view which was dissimilar to the opinion of the committee as follows: It is feasible that the flow velocity of the main flow downstream from the orifice with the smaller cross-section was appreciably higher just as depicted in Fig. (**6-19**), and consequently the intensity of turbulence also higher than the fluid flow on the upstream side. However, the effect of the turbulence in the main flow must not reach the pipe wall surface where the pattern appeared, because the wall surface downstream from the orifice faced to the secondary flow of lower flow velocity or the stagnant watershed. Therefore, the wall thinning process seemed to be in the condition which was similar to electrolysis or etching submerged in a standstill liquid. Then, it would be appropriate to consider that the observed fish-scale pattern (Fig. (**6-29**)) must represent the inherent microstructure of the carbon steel surface rather than the influence of shear force or turbulence. While it is difficult to know the dimensions of the pattern since no scale was provided in the micrograph of Fig. (**6-29**), in other case than Mihama, the diameter of each fish-scale in the similar characteristic pattern was 1-2 mm. As the carbon steel microstructure in proportion to this size, the crystal grain of the austenite which appears in high temperature of steel manufacturing process is considered [7].

The ground for the above mentioned authors' opinion is the wall thinning case shown in Fig. (**6-26**): In the inside wall of the pipe of issue, the small fish-scale pattern can be recognized (Fig. (**6-27**)). In Fig. (**6-30**), the magnification of the pattern is shown: the size (average diameter) of four fish-scales in visual field of the SEM micrograph is a few mm.

The SEM micrograph shown in Fig. (**6-31**) is the magnification of the central part of the fish-scale which is located at the center of Fig. (**6-30**): many black planes edged with the white undefined material can be recognized. In Fig. (**6-32**), the comparatively large black plane in center left side of the SEM micrograph in Fig. (**6-31**) is magnified.

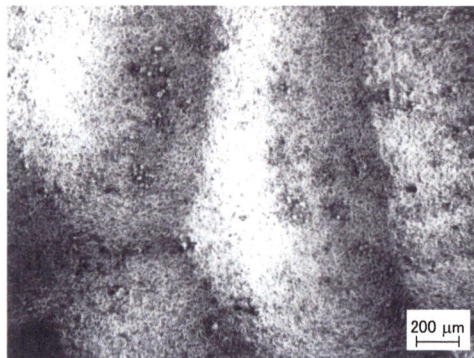

Fig. (6-30). Four fish-scales in fish-scale pattern.

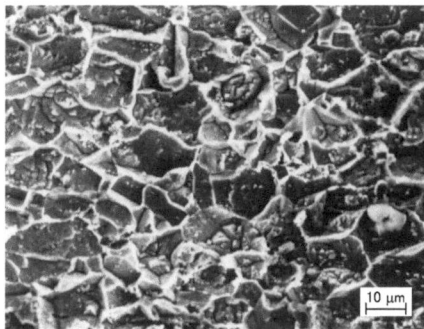

Fig. (6-31). Magnification of the central part of the fish-scale in Fig. (**6-30**).

The surface of concerned plane as well as those comparatively small planes surrounding the large one is flat and smooth. In addition, it can be observed that the direction of each plane is fixed to a common direction regardless the size of plane. This fact may be proved more clearly with the lots of black planes in the micrograph of Fig. (**6-31**), because, otherwise, planes would appear black, white or gray due to the irregular reflection of electron beam from the surfaces. These black planes are presumed to be one of the crystal planes of ferrite, that is, {100} or {110} which is of the lowest and second lowest dissolution rate. There is no such situation in the microstructure of carbon steel where the crystal planes are arranged uniformly like this, except for the vicinity of the old austenite grain boundary where the epitaxial crystal growth used to occur. Thus, the fish-scale pattern in Fig. (**6-29**) was concluded to have been derived from the old austenite grain boundary.

Fig. (6-32). Magnification of relatively large black plane in center left side in the SEM micrograph of Fig. (**6-31**).

Lastly, it should here be strongly emphasized that the fish-scale pattern mentioned above has to be completely unrelated to "scallop appearance" which used to appear in the erosion-corrosion damaged surface of carbon steel pipe as shown in Fig. (**7-3**) of the next chapter.

REFERENCES

[1]   Murakami M, Yabuki A, Matsumura M. Is the Damage to Pure Copper Piping an Erosion-Corrosion in Nature? Zairyo-to-Kankyo 2004; 53: 440-445.

[2]   Suezawa Y, Shinohara, T. Dependence of Corrosion Rate of Metals upon the Structure and Finish-condition of their Surface-layer. J. Mater. Sci. Soc. Jpn. 1968; 5: 160-173.

[3]   http://www. nisa. meti. go. jp/.

[4]   Furuya Y, Murakami M, Yamada Y. Ryuhtai-kohgaku. Tokyo: Asakura-shoten 1967; pp. 68-73.

[5]   Umemura F, Kawamoto T. Stress Corrosion Cracking of Carbon Steel and Low Alloy Steel in High Temperature Water. Boshoku-Gijutsu (presently Zairyo-to-Kankyo) 1981; 30: 276-281.

[6]   Bouvier O. de, Bouchacourt M, Fruzzetti K. Redox conditions effect on flow accelerated corrosion: influence of hydrazine and oxygen. In: Proceedings of the International Conference of Water Chemistry of Nuclear Reactor Systems; 2002: Avignon, France: pp. 22-26.

[7]   Matsumura M. Features and Mechanism of Unusual Wall Thinning (Erosion-Corrosion) of the Carbon Steel Pipe Carrying Hot Pure Water. Zairyo-to-Kankyo 2007; 56: 187-195.

CHAPTER 7

Prevention of Erosion-Corrosion in the Field

Abstract: In planning appropriate and effective countermeasures to protect equipment and machines against the attack of erosion-corrosion, it is indispensable to know the true nature of the attack, so that erosion-corrosion is defined as flow induced localized corrosion in the morphology, and as flow induced macro-cell corrosion in the mechanism. The water chemistry controls mainly the chemical property of the environmental liquids such as pH or dissolved oxygen content in order to prevent uniform corrosion, and accordingly this countermeasure is judged rather not suitable for preventing erosion-corrosion which is categorized into localized corrosion. Selection of material depends not only on the corrosiveness of the environmental liquid but also the circumstances of industry. The cathodic protection, which has been long utilized for preventing the electrolytic corrosion of underground buried pipelines, has a high possibility as a countermeasure to mitigate erosion-corrosion damage because it can be applied to localized corrosion. The elimination of those irregularities in fluid flow, which induce the generation of the macro-cell, is recommended most as a successful countermeasure against erosion-corrosion, and the detailed examples are given. Lastly, methods for monitoring the inception as well as the activity of macro-cell are described. Following the concept of "leak before break" a method to prevent the rupture of the pipeline is also shown.

Keywords: Water chemistry, dissolved oxygen content, cathodic protection, anodic protection, erosion-corrosion, monitoring, macro-cell, electrolytic corrosion, inhibitor, scallop pattern, fish scale pattern, magnetite, hematite.

1. DEFINITION OF EROSION-CORROSION AND THE PREVENTION METHODOLOGY

In selecting the countermeasures to protect the machines and equipment from the attack of erosion-corrosion, it is indispensable to recognize the true color, or, the mechanism of the attack, because the effective and stern countermeasures totally depend on it. The mechanism of erosion-corrosion has been described in detail in preceding chapters, which may be summarized in the definition terminology for erosion-corrosion: "flow induced macro-cell corrosion (FIMC)". This may closely resemble the term "flow induced localized corrosion (FILC)" which had already been proposed by Heitz, DECHEMA, in 1990. The difference between these terms is that the former referred the mechanism of erosion-corrosion, and the latter the mode (morphology) of it. However, later, he regretfully changed his claim from FILC to FLC, "flow induced corrosion" [1]. He might presumably have intended to expand the range of application for the term: the exclusion of the word "localized" from the definition term apparently means the elimination of the distinction between uniform corrosion and localized corrosion. Contrary to his intention, the exclusion caused a truly serious influence, at least, on the selection of the countermeasures for erosion-corrosion because they are quite different depending on the mode of corrosion attack, uniform or localized, as it is described in the followings.

The water chemistry, which is nowadays in the wide use as the countermeasure to protect the pipelines of carbon steel from erosion-corrosion, is, from the view point of corrosion theories, to be applied to the uniform corrosion. Therefore, it is not the exaggeration that this corrosion protection method is ineffective for every type of erosion-corrosion including FAC because it is a localized corrosion. Upgrading of the material is not appropriate either, because this also assumes rather uniform corrosion. In comparison with these measures mentioned above, the electrolytic protection, the cathodic and anodic protection, can be applied to uniform corrosion as well as localized corrosion. This methodology has long been applied for protecting the underground metallic pipelines (pipeline of cast iron for the tap water service, oil transporting carbon steel pipelines *etc.*) from the stray electric current s which leaked from the rails of DC electric power train, but there is no example yet in which it be applied to protect pure water carrying pipelines not laid in underground but hanged overhead. Nevertheless, the possibility of success cannot be avoided because the electrolytic corrosion of underground metallic pipelines is a typical localized corrosion and accordingly categorized in macro-cell corrosion similarly in the case of erosion-corrosion which occurred in the pipeline of carbon steel at elevated temperatures. Other corrosion protection methods, such as painting and plating are to be excluded because of difficulty in the execution under the circumstances where erosion-corrosion may occur. In conclusion, the suitable and right countermeasures against erosion-corrosion lie not in the field of the electrochemistry but rather in the field of the fluid dynamics because favorable effect of improvements in the flow conditions is expected for avoiding the inception of macro-cell of corrosion on the pipe wall surface.

Assuming the case, where the countermeasures mentioned above failed in preventing it, monitoring systems for detecting the inception of macro-cell of corrosion are required. A sort of fail-safe system should be also attached in order to avoid at least the rupture of the pipeline in any case.

2. WATER CHEMISTRY

2.1. Boiler Feed Water Treatments

The water chemistry is a technique which intends to suppress the corrosiveness of working liquid through controlling the chemical as well as physical property of the liquid such as pH, the content of dissolved oxygen, and the conductivity. One of the application examples of the methodology is the boiler feed water treatments such as AVT, CWT and NWT which are listed on Table **5-2** in Chapter 5. And, exactly in the same chapter, it was described that erosion-corrosion regretfully occurred on the jet-in-slit specimens of carbon steel in those boiler waters mentioned above.

In the boilers of huge capacity, such as at the nuclear or thermal power plants, the steam temperature is elevated higher than the critical temperature of water in order to achieve a higher thermal efficiency. Since, the distinction between liquid and vapor disappears in the supercritical state, those boilers are not equipped with the steam-liquid separation drums, and the water supplied is totally evaporated into the vapor. Therefore, only volatile reagents are used to control the level of pH or DO in supply water, so that they do not deposit on the inside wall surface of the heat conducting tubes: this is the All-Volatile Treatment (AVT). In the meantime, the boilers of comparatively smaller scale are operated below the critical temperature, and are equipped with the steam separation drums. Since, the reagents remained in the liquid phase and condensed too much are discharged by the drain blow from the drum to the outside of the system, inhibitors can be utilized to assist stable protective films to form on the metal surface. This is also categorized in water chemistry, but inhibitors are still not always suitable for preventing erosion-corrosion.

Irrespective of the capacity of boiler, the boiler feed water is made from the de-ionized water in order to avoid the formation of scale on the pipe wall surface or the generation of the sludge in the drum. The expectation that any electrochemical corrosion might be not possible in this water with the high resistivity (ca. 1 MΩ cm) is granted in the stagnant water but betrayed in the flowing water, the reason for which is explained as follows: even in the pure water at pH 7 the certain amount of ions are contained: H^+ and OH^- ions of 10^{-7} mol L^{-1} each. And, even though they scarcely carry the electric charge through migration, which is the main reason for the high electric resistivity, the ions are, accordingly the electric charges are carried by the convection in the flowing water, resulting in more or less the electrochemical corrosion: ions in flowing fluid are transferred through diffusion, migration as well as convection just as the famous Nernst-Planck equation clams. When a macro-cell is consisted in addition to it, corrosion advances at substantially high rate as the consequence of the macro-cell current effect as well as the surface area ratio effect [2].

In conclusion, as long as the uniform corrosion in stagnant water is considered, the water chemistry is an excellent methodology which is useful not only for the prevention but also for controlling the corrosion, with comparably lower costs and the easiness in execution. Nevertheless, it would not demonstrate sufficient performance to erosion-corrosion since it can not prevent the inception of macro-cell of corrosion in the flowing water. Inhibitors can be, what is worse, even the indirect cause of erosion-corrosion as illustrated in the following case studies.

2.2. Case Studies of Water Chemistry Use

The first case is successful use of water chemistry. A container of Type SUS316L stainless steel was used in the environment of condensed process water containing NH_3 and CO_2, and it was revealed that the process water had general corrosiveness of near 1 mm y^{-1}, so that a corrosion prevention method, that is, the water chemistry was examined as follows. First, the container was filled with air-blown water. Then, the condensed process water was injected to the air-blown water without substantial change in the potential of the container. The potential of the container was still maintained noble even after the air blowing was stopped (Fig. (**7-1**)). Thus, the air blow, that is, the fluent supply of dissolved oxygen was utilized successfully in prevention of uniform corrosion on the container wall surface [3].

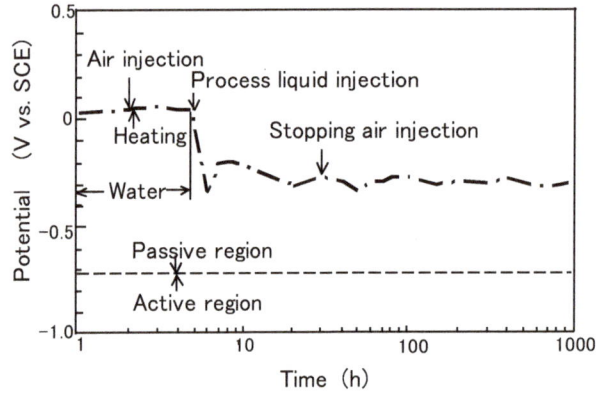

Fig. (7-1). Successful application of water chemistry to uniform corrosion: variation of container potential made of stainless steel [3].

The second is the case of erosion-corrosion in which the inhibitor was the indirect cause of it. The inhibitor (phosphoric acid type) was being continuously supplied through the T jointed branch tube (15A) to the main line (50A) which was circulating the boiler feed water.

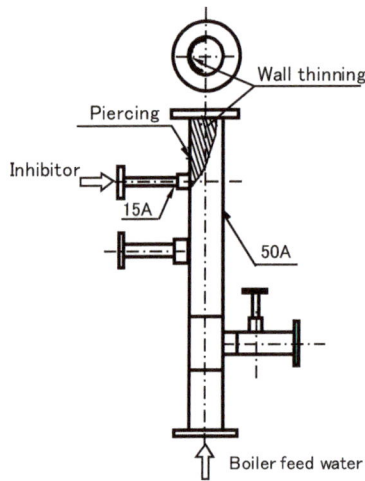

Fig. (7-2). Details of concerned piping of carbon steel.

Fig. (7-3). Scallop patterns appeared on the pipe wall surface at the downstream from the inhibitor inlet.

Soon, a small hole pierced the main pipeline through the wall of branch tube side at the location downstream from it (Fig. (7-2)). The so-called scalloped appearance was recognized, in which tongue shaped patterns were arranged uniformly in the flow direction on the internal surface of the pipe, and the piercing was found on the bottom of the deepest tongue (Fig. (7-3)). The generation process of the open hole was estimated as follows.

1.   A deflecting jet was formed as the consequence of the correlation between the flow conditions in the branch tube and the main pipe such as radius ratio, flow velocity ratio, *etc.* (Fig. (7-4)).

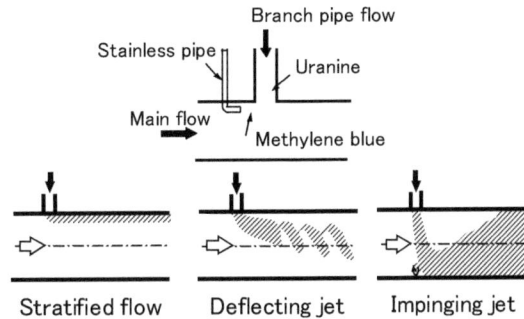

Fig. (7-4). Characteristic flows in T junction of piping system [4].

2.   Since the deflecting jet from the branch tube played the role of the obstruction to the main flow in the pipeline, the longitudinal vortexes, that is, the arched vortexes along edges of the jet as well as lateral vortexes, that is, the wake vortexes were generated downstream from the jet on the pipe wall surface [4]. And, the latter covered the wall surface spot-wise but contentiously (Fig. (7-5)).

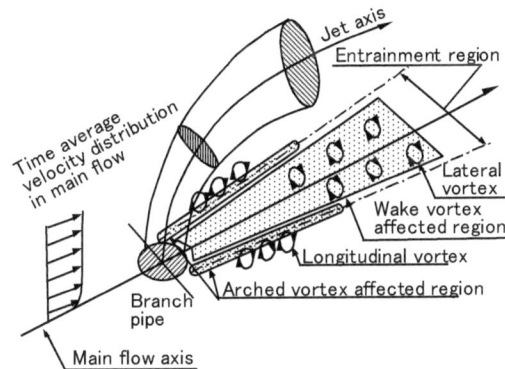

Fig. (7-5). Generation of vortexes at a T junction of pipes [4].

3.   On the wall surface under the vortex, the inhibitor content was much lower than that of outside because it had been consumed soon but the supply from the main flow was rather meager (Fig. (7-6)).

4.   The corrosion rate was higher under the vortexes due to the want of inhibitor. On the other hand, the corrosion rate on the pipe wall surface outside of the vortex was comparatively lower due to the inhibitor contained sufficiently in the main flow. The macro-cell of corrosion was formed for the surfaces where the corrosion rates were different. The wall thinning proceeded at the macro-anode, on the wall surface under the vortexes, with high rate resulting in the piercing.

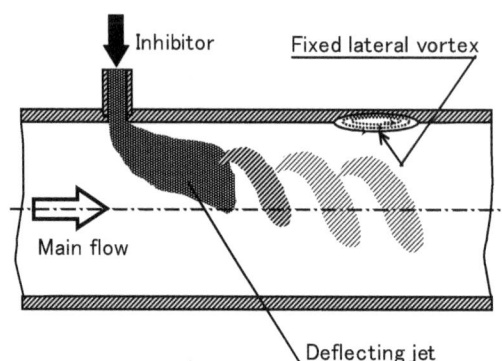

Fig. (7-6). Fixed vortexes generated downstream from a deflecting jet.

3. SELECTION AND UPGRADING OF MATERIALS

3.1. Material Selection for Components of Facilities

Upgrading of materials may be necessary for the corrosion damaged equipment which engages with the large amount of once through seawater or tap water where the water chemistry cannot be applied. And, the degree of which the corrosion resistance of material is emphasized in materials selection is totally dependent on the corrosiveness of water and the circumstance of the industry as well. As an example, the following are taken up: valves for the tap water (in connection with Chap. 5) and heat conducting tubes of seawater heat exchangers in the thermal and nuclear power plants.

Material Selection for Tap Water Valves: For the raw materials of the domestic tap water valves, the anti-corrosion performance is a must, and in this aspect bronzes (copper-tin alloys) have satisfactory corrosion resistance, so that most valves for the tap water use in Japan are made of bronzes. However, the industry was not satisfied with the price of bronzes because the valves are great in number, and, therefore, even a slight change in the material cost may result in enormous amount of money. Being connected with this, the material cost consists of the purchase cost and the manufacturing cost, which makes the material selection more complicated.

Brasses (copper-zinc alloys) are inferior to bronzes in corrosion durability due to the dezincification corrosion, but the market price of zinc is nearly one fourth of the tin price. Then, the industry tried to supplement the defect of brasses, and dezincification proof brasses were developed through alloying basses with a small amount of the third elements such as arsenic and others.

Besides bronzes and brasses, there are naturally various copper based alloys with excellent corrosion durability, but they have some consentient reasons for not to be adopted for water taps. For example, aluminum is cheaper than zinc. Hence aluminum bronze (copper-aluminum alloy) appears most advantageous in the cost of raw material. At the same time, it possesses excellent mechanical (strength *etc.*) as well as chemical properties including corrosion resistance. Nevertheless, it is seldom used for water taps because the machining cost is high, and also because qualified skill is requested at the manufacturing stage since aluminum is easily oxidized at elevated temperatures.

Thus, the candidates for the raw materials of valves for the tap water were, as listed in Table **5-1**, Chapter 5, five bronzes, three brasses and two dezincification proof brasses. It should be recognized that the intention of the corrosion tests for selecting the materials was not upgrading but rather downgrading of materials.

Material Selection for Heat Conducting Tubes in Seawater Heat Exchangers: The thermal and nuclear power plants are in a very severe environment because of their so large scales: when the operation is stopped by accident, so many customers possibly suffer so serious damage. In Japan, the possibility of accident in power stations might be higher than other countries because most power stations are utilizing the seawater for the coolant of the heat exchangers which condense the steam into water after the drive of the power generating turbine: the seawater has the merit of inexhaustible and no charge, but at the same time the demerit of the troublesome bio-fouling and intense corrosiveness. Nevertheless, absolutely no leakage due to corrosion is allowable for the heat conducting tubes in seawater heat exchanger as described above. In order to respond to this sever demand, a copper base alloy (Cu-20Zn-2Al-0. 35Si), named as Albrac, with excellent seawater resistance was developed in Japan so early as in 1932. However, the industry was not completely satisfied with its performance, so that the improvement on the material of the condenser tube has been kept in the following succession: Alum-brass (Cu-20Zn-2Al-0. 6Ni), AP bronze (Cu-8Sn-1Al-0. 2Si), 10% Cupronickel (Cu-10Ni), 30% Cupronickel (Cu-30Ni), and today almost all of condenser tubes in power stations in Japan are of titanium alloys.

3.2. Upgrading of Materials and Macro-Cell Corrosion

The followings are the cases where erosion-corrosion has been successfully controlled by upgrading of materials: the damage of localized graphitization corrosion near the shaft hole in the casing of seawater pump was successfully avoided through upgrading the material from a gray cast iron to a 1% Cr cast iron (Table **7-1**). Another case is also shown in the table which is of the carbon steel pipe carrying the pure water at 145 °C: through upgrading the material from the carbon steel to a low alloyed steel (1Cr-0. 5Mo steel), the pipe wall thinning rate was reduced down to one tenth of the carbon steel.

Table 7-1. Performance of materials

Seawater pump casing of cast iron	
Gray cast iron (FC20):	Unusable
1% Cr cast iron (ES 51F, Ebara Co):	Usable
Steel pipeline carrying pure water at 145 °C	
Carbon steel (STPT38S):	0. 78 mm/y
1Cr 0. 5Mo Steel (STPA22S):	0. 07 mm/y

In the case of cast iron, the formation of macro-cell was still recognized in the low alloyed cast iron. Nevertheless, this countermeasure was approved. In contrast, in the case of the carbon steel, it can be estimated that the macro-cell was not formed in the low alloyed steel any more. Nevertheless, the author cannot agree with the upgrading of this case. The reasons are given below.

Protecting Seawater Pumps against Erosion-Corrosion: Kitajima *et al.* investigated how the addition of the alloying element, chromium, in the gray cast iron had improved the resistance to erosion-corrosion [5]. They cut the casing of the pump down into small fractions (as shown by dotted lines in Fig. (**1-12**), Chapter 1), then put leading wires on each piece, and bonded them together with epoxy resin to rebuild the casing as well as the pump. While pumping seawater, they measured by using a zero-shunt ammeter the macro-cell current which flowed into the fractions near the shaft hole.

The results of the measurements are given in Fig. (**7-7**): the solid line for the gray cast iron casing; the dotted line for the 1% Cr cast iron casing. It can be clearly recognized that the current flowing in the 1% Cr cast iron casing was as small as one tenth of the gray cast iron, and accordingly that the upgrading of cast iron was quite effective in reducing the dissolution of ferrous ions. The reason of this favorable effect of chromium is assumed that this alloying element must have assisted the formation of a dense residue layer on the surface which must substantially prevent the migration of Fe^{2+} ions.

At the same time, the behavior of the currents in Fig. (**7-7**) proves the formation of macro-cell in the 1%Cr cast iron casing. The rationale is illustrated in Fig. (**7-8**), where the relationships between the dissolution rates of ferrous ions and time for the different surfaces in the casings, one near the shaft hole and the other of the surroundings, is shown: each current in Fig. (**7-7**) corresponds to the difference between the dissolution rates of ferrous ions at those surfaces. And this difference is nothing but the macro-cell current.

The formation of macro-cell of corrosion is easily recognized for the casing of gray cast iron as well as that of 1%Cr cast iron.

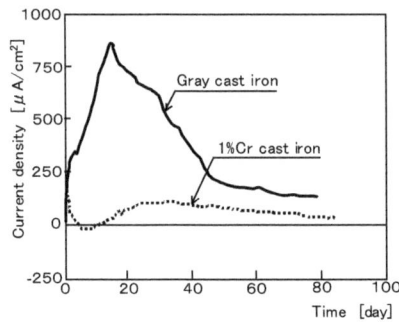

Fig. (7-7). Currents flowing into the fraction near the shaft hole [5].

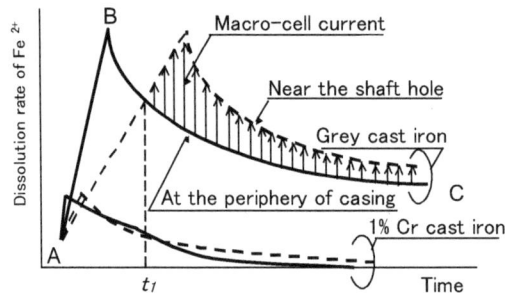

Fig. (7-8). Rationale for the formation of macro-cell of corrosion.

In protecting seawater pumps from erosion-corrosion, the water chemistry cannot be applied, the flow velocity on the surface of the casing cannot be altered either. The only way left is the upgrading of the material in order to reduce the graphitization corrosion rate and the macro-cell corrosion rate accordingly.

Protecting Carbon Steel Pipes from Erosion-Corrosion: The erosion-corrosion case of the carbon steel pipe described in Table **7-1** is one of the three cases listed on Table **6-1** in Chapter 6: the wall thinning due to erosion-corrosion caused the carbon steel pipeline, which had been transporting the pure, high-temperature water, to rupture at the downstream from the flow nozzle. The generation mechanism of this severe wall thinning was explained as follows. A stagnation watershed was inevitably formed at the downstream from the nozzle flow meter installed in the pipeline. The supply of dissolved oxygen from the main stream to the pipe wall surface under the stagnation watershed was rather meager than that to the pipe wall surface facing directly to the main stream (Fig. (**6-19**)), which brought about the situation that, in the critical temperature range of active/passive transition for corrosion state, the wall surface under the stagnant watershed was still active while the wall surface facing to the mainstream was already in the passive state. Owing to the great difference in the corrosion rates between those surfaces a macro-cell of corrosion was formed and the anodic dissolution at the macro-anode proceeded at a higher rate due to the macro-cell current effect as well as the surface area ratio effect, resulting in the serious wall thinning.

This process of macro-cell formation in the carbon steel pipes described above is illustrated in the corrosion rate *vs.* temperature diagram in Fig. (**7-9(a)**).

The same for the low alloyed steel is given just under the diagram (Fig. **7-9(b)**), where the critical temperature of the active/passive transition moved to lower temperature side, and at the wall temperature, that is t_w, the wall surface under the stagnant watershed as well as that facing to the main stream are complete passive state, resulting no difference in corrosion rate between them, and accordingly no formation of macro-cell. It must be, however, recognized that in Fig. (**7-9 (b)**), the critical temperature for active/passive transition still exists. Suppose another nozzle installed downstream from the original one, and the temperature of a pipe wall surface which located downstream from the nozzle were t_{w2} as indicated in the diagram, the formation of macro-cell would not be avoided. This is the reason why the author denied the usefulness of the upgrading of material in this case.

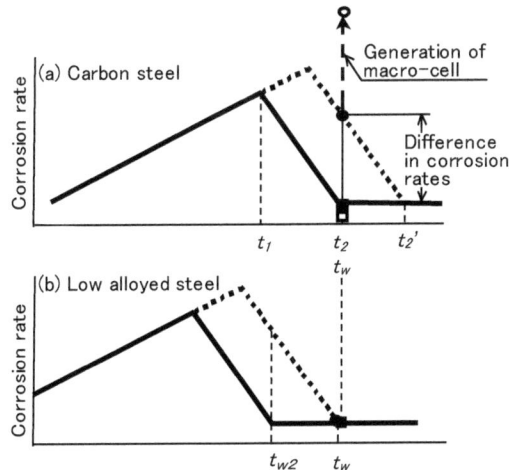

Fig. (7-9). Formation of macro-cell of corrosion in carbon steel pipe and low alloyed steel pipe.

When the pipes of carbon steel were upgraded to those of stainless steels which are in passive state in the entire temperature ranges, the galvanic corrosion would occur in the neighboring carbon steel pipes. The upgrading the material of the entire pipelines to stainless steels is ranked in the position most separated from the approval of the industry.

4. CATHODIC AND ANODIC PROTECTION

4.1. Theories of Cathodic Protection

The technique of cathodic protection has been employed long before the occurrence of FAC problems as the measure preventing electrolytic corrosion or stray-current electrolysis, which is defined as accelerated corrosion originated through DC stray currents, of underground metallic structures such as pipelines and storage tanks. Important is that the electrolytic corrosion is categorized as a localized corrosion in the mode, and accordingly as a macro-cell corrosion in the mechanism. The cathodic protection may be, therefore, advantageous as a countermeasure to the localized corrosion of carbon steel pipelines at elevated temperatures, that is, FAC, as compared with the water chemistry which is rather suitable for preventing uniform corrosion. Besides, this method can be comparatively simply applied, even after the corrosion is generated. In this point, it resembles the water chemistry.

Principle of Cathodic Protection: In the text books of corrosion, the principle of cathodic protection is comprehensively illustrated by using the schematic drawings of experimental set-up in laboratory (Fig. (**7-10**)) and an Evans diagram (Fig. (**7-11**)).

When the direct electric current is impressed from the auxiliary electrode of inert material such as graphite or platinum through the environmental solution to the specimen which is corroding with the rate of i_C at the corrosion potential, E_C, the anodic current is reduced from i_C down to i_I, and at the same time, the level of potential of the specimen is lowered from E_C down to E_I. Further application of the protection current puts the specimen potential down to the level of equilibrium potential of the concerned electrode reaction, E_0, and at this potential the anodic current is reduced to substantially zero. Thus, the specimen is completely protected from corrosion. Here, the micro-cell corrosion, the uniform corrosion therefore, is assumed for the specimen, because there is only a single anodic polarization curve in the Evans diagram.

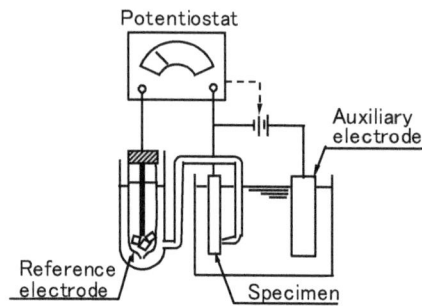

Fig. (7-10). Experimental set-up to conduct cathodic protection in laboratory.

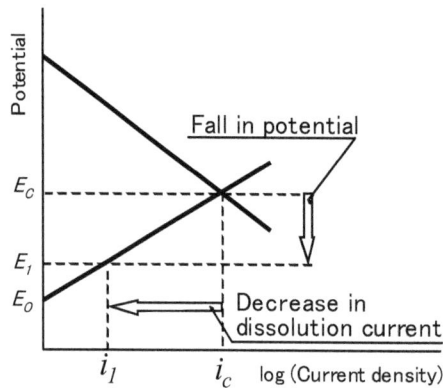

Fig. (7-11). Evans diagram to illustrate the principle of cathodic protection.

Mechanism of Electrolytic Corrosion of Underground Pipelines: The scheme of electrolytic corrosion of underground metallic pipeline caused by the leakage current from electric railway in which the steel rails are used for the current return to the substation is illustrated in Fig. (**7-12**): a direct current is supplied from the substation through the trolley wire to the train; it powers the motor of train, and then returns through the tier and the rail back to the substation. At the location A, a part of the current leaks out of the rail to the earth, and then still flows through the soil back to the substation. This current which is flowing through the soil of high resistivity constructs the considerably intense electric field in the soil where the potential is at maximum in location A, and descending along the rail down to the minimum in location B which is most closed to the substation. At location A, ions deposit on the surface of the pipeline resulting in the current flow in it. This current flows through the pipeline in the direction approaching the substation. At the location B, which is sited most close to the substation it flows out of the pipeline and returns through the soil back to the substation, accompanying large corrosion damage in the wall of the pipeline. The location N sited between the location A and B is called as neutral point where no current flows into the pipeline or out of the pipeline either.

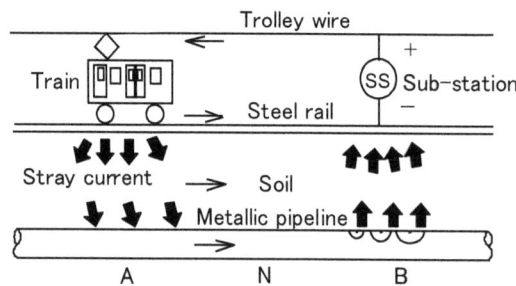

Fig. (7-12). Scheme of electrolytic corrosion in underground metallic pipeline.

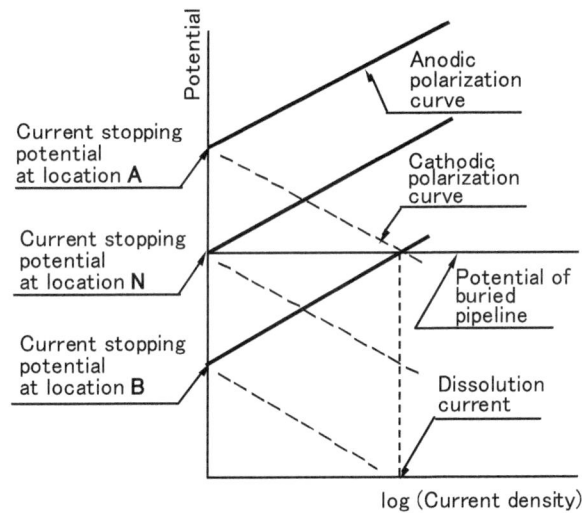

Fig. (7-13). Evans diagram to rationalize the scheme of electrolytic corrosion shown in Fig. **(7-12)**.

The scheme above is rationalized with the Evans diagram in Fig. **(7-13)**, where the paired polarization curves, which are similar to each other, are given for each of the locations cited in Fig. **(7-12)**, provided that no other reaction but the reduction and oxidation of iron is considered for the simplicity. This might appear that the equilibrium potential of an electrode reaction be dependent on the location of the metal surface, which cannot be recognized in the usual cases where a usual current is impressed in a usual range of extent just as the case of the laboratory set-up in Fig. **(7-10)**. Nevertheless, in the case of a high voltage (thousands volt) in the environment of higher specific resistance, the soil, an intense electric field arises, where the influence of electric field potential or so-called IR drop on the equilibrium potential cannot be regarded uniform: in other wards, in an intense electric field, the cathodic current and anodic current cannot be equal at the equilibrium potential, but either is predominant: they are equal at a potential somewhat deviated from the equilibrium potential, and the extent of the deviation depends on the extent of field potential, which means the deviation changes by the place in the soil. This situation is quite inconvenient in discussing the mechanism of electrolytic corrosion, so that a new parameter, E_{ST}, is introduced, which is defined as "current stopping potential" and given by the following equation.

$$E_{ST} = E_0 + P \qquad\qquad\qquad\qquad\qquad (7\text{-}1)$$

$$E_0 = E_0^0 + (RT/nF) \times \ln a \qquad\qquad\qquad\qquad\qquad (7\text{-}2)$$

$$\Delta P = I \times R \qquad\qquad\qquad\qquad\qquad (7\text{-}3)$$

$E_{ST,}$ current stopping potential;

E_0, equilibrium potential;

E_0^0, standard equilibrium potential;

a, activity of ion;

P, potential of electric field;

I, stray current;

R, resistance of soil

According to Eq. (7-1), the potential of electric field as well as the chemical potential (ion activity) functions as a driving force which push the current from the environment, soil, to the metal. The current stopping potential varies place to place along the underground pipeline resulting in multiple anodic and cathodic polarization curves. Thus, the electrolytic corrosion of underground metallic pipeline described above is categorized as localized corrosion in the mode, and as macro-cell corrosion in the mechanism since multiple anodic polarization curves are concerned.

Cathodic Protection of Underground Pipelines: Fig. (**7-14**) shows the scheme of cathodically protected pipeline with deeply and dispersedly buried anodes (auxiliary electrodes), where the DC protection current flows right-angled and uniformly at each part of the buried pipeline. Consequently in this case, the relative distribution of potential in the electric field, and accordingly that of current stopping potential at each location of the pipeline is not changed. Instead, the potential of the pipeline is lowered, and the electrolytic corrosion damage at the location B is reduced. This mechanism is identical with the principle of the cathodic protection for uniform corrosion (micro-cell corrosion) shown in Fig. (**7-11**).

Fig. (**7-15**) shows the scheme in which an anode is installed in the vicinity of location B where the damage of electrolytic corrosion is concentrated. Since the protection current is substantially concentrates in the location of B, the potential of the electric field in the place rises, and accordingly, the current stopping potential of the pipeline at the location rises, which moves the intersection point between the potential line and the anodic polarization line, that is, the corrosion current, toward the lower current side. Thus, the electrolytic corrosion damage of this place is reduced.

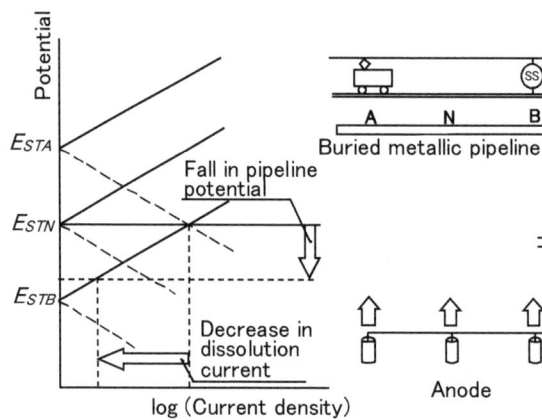

Fig. (7-14). Evans diagram rationalizing cathodic protection with dispersed anodes.

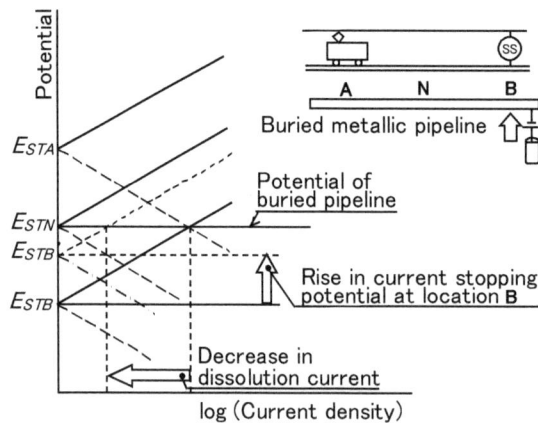

Fig. (7-15). Evans diagram rationalizing cathodic protection with a single anode.

In conclusion, the damage of electrolytic corrosion, which is a localized corrosion in the mode and macro-cell corrosion in the mechanism, is reduced through two principles of cathodic protection: lowering the pipeline potential and raising the current stopping potential.

The mechanism of cathodic protection described above makes, regretfully, the application of this protection method hesitate to be applied to the localized corrosion of carbon steel in high-temperature water (for example, Fig. (**6-24**)). This is because this sort of localized corrosion is of an active/passive type macro-cell corrosion, and accordingly there is the possibility that the passive state of pipe wall be demoted to an active state due to the collapse of passivity, in any way the current stopping potential ascends or the pipe wall potential descends with the impress of cathodic current, resulting in a rise of corrosion rate. Instead, in this case, the state of the tube wall at downstream from the orifice should be rather promoted from active to passive, then, the macro-cell must be decomposed and the corrosion rate remarkably lowered.

4.2. Anodic Protection

In contrast to the cathodic protection, the application of anodic protection is limited to nickel, iron, chromium, titanium and their alloys which are given the active/passive shift, since the technique is based on the formation of passive film on the surface. The erosion-corrosion occurred on the carbon steel in the pure water at elevated temperatures is optimum as the object of anodic protection, since it is macro-cell corrosion originated through the active/passive shift.

In order to prevent the localized corrosion in the pipe wall facing the stagnation watershed at the downstream from orifice, the anodic current is to be impressed from the macro-anode, that is, the pipe wall surface, through the water to a cathode (Fig. (**7-16**)). Then, it becomes passive similarly to the wall surfaces facing the main stream, the difference in corrosion rate between the wall surfaces as well as the macro-cell being disappeared, as illustrated with the Evans diagram in Fig. (**7-17**). This is correspondent to the case in which current stopping potential is controlled by the cathodic protection described above, because in this case, not potential of the pipeline but polarization behavior of the macro-anode is controlled.

The orifice plate of carbon steel may be conveniently utilized for the cathode or the auxiliary electrode, since it is protected with the cathodic protection current, which is advantageous in making the electrolytic protection system concise as well as in completing without any change in the flow system and the conditions. You should consult on practical operation conditions such as the potential and the duration of anodic current impression, which may be necessary for the completion of anodic protection, at Fig. (**5-25**) and the similar operation depicted in Fig. (**7-1**).

In conclusion, the anodic protection is most suitable as the posterior countermeasure. In other words, it is not a prevention measure but rather a mitigation to be conducted after the rise of erosion-corrosion damage.

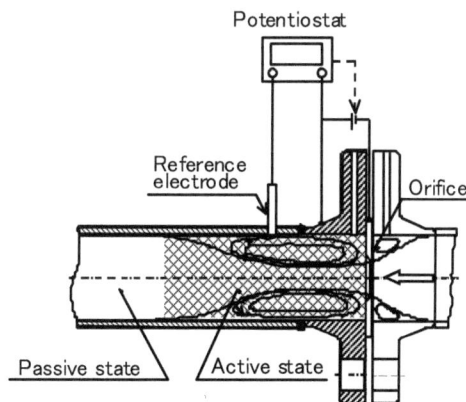

Fig. (7-16). Anodic protection to prevent the localized corrosion in the pipe wall faced to stagnation watershed.

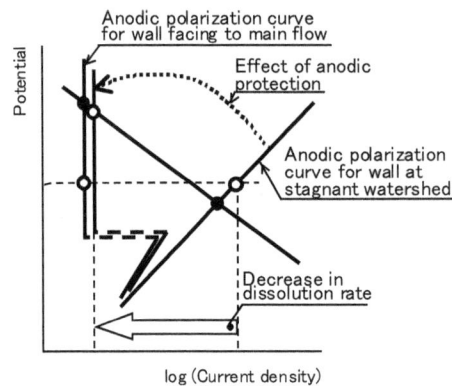

Fig. (7-17). Evans diagram rationalizing the scheme of anodic protection in Fig. **(7-16)**.

5. ELIMINATION OF IRREGULARITIES FROM FLUID FLOW SYSTEMS

5.1. Elimination of Irregularities in Flow Conditions

As it was described at the beginning of this chapter, one of definitions for erosion-corrosion is "flow induced localized corrosion (FILC)". The essential of the countermeasures against erosion-corrosion is, therefore, not to permit the flow to induce erosion-corrosion. Another definition of erosion-corrosion is "flow induced macro-cell corrosion (FIMC)". A macro-cell of corrosion is induced when difference is originated in the anodic dissolution rate on a metal surface through the difference in flow conditions of the environmental liquid. So that, the first step in the prevention of erosion-corrosion is to eliminate the differences in flow velocity. A typical successful example is the mitigation of erosion-corrosion rates with the guide vane in the bend of pure copper as it was described in Section 1 of Chapter 6. Other possible applications of the countermeasure along this idea are as follows.

Elimination of Stagnant Watershed at Downstream from Orifice: As shown in Table **6-1** and Figs **(6-13)** to **(6-19)**, serious wall thinning leading to the rupture of the carbon steel pipeline carrying pure water at elevated temperatures (FAC) occurred so often at the downstream from the orifice or nozzle flow meters. These were attributed to the formation of the macro-cell of corrosion which was resulted from the difference in active/passive state between the wall surface under the stagnant watershed (Fig. **(6-19)**) and the surface facing to the main stream. The countermeasure derived from the rationale described above is elimination of the stagnant watershed, which may be achieved with the replacement of the orifice or nozzle with a flow meter without the stagnant watershed: the total pressure tube (Pitot tube) shown in Fig. **(7-18)** is the most possible example because there is no necessity of the remodeling of attached facilities for pressure measurements.

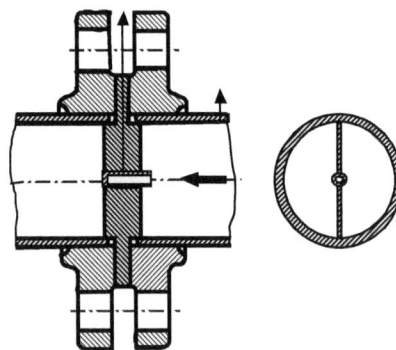

Fig. (7-18). Total pressure tube (Pitot tube) in place of orifice.

Elimination of Dead Water at the Entrance of Pipes and Fittings: The stagnant watershed or the dead water is also found at the entrance of pipes and fittings. The famous "guillotine" rupture of the pipeline of carbon steel in the nuclear power plant at Surry, Virginia, assumingly corresponds to this case [6]: among the lines that were running in parallel, pipeline A (18" in nominal diameter) was connected with a 24" header *via* a branch pipe of the same size (18"), as shown in Fig. (**7-19**).

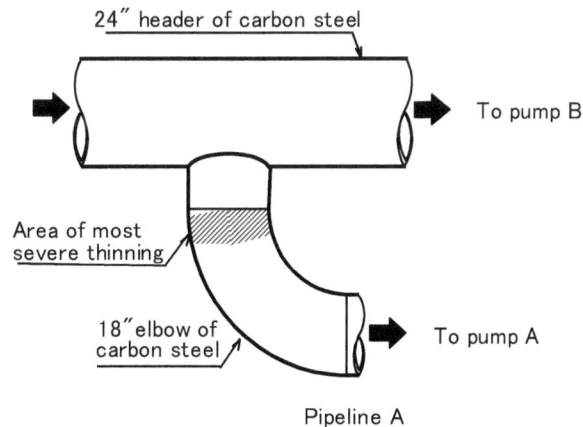

Fig. (7-19). Elbow in pipeline A connected with header *via* branch pipe [6].

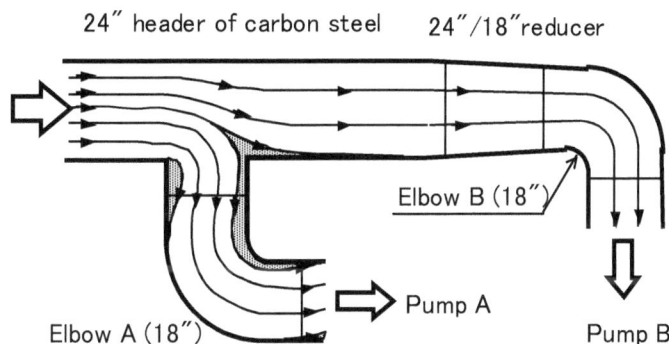

Fig. (7-20). Sketch of flow pattern in elbows connected with the same header in different ways.

In contrast, the elbow in pipeline B (18") was connected with the header *via* a 24"/18" reducer as shown in Fig. (**7-20**). The simple sketch of stream lines in the pipeline clearly indicates the difference in the conditions of water flow between the two elbows: the inside surface of elbow A must be covered with the stagnant watershed that had developed at the T junction, while there must be a smooth flow along the entire surface of elbow B. This must be the reason for that elbow B was free from that type of wall thinning. In contrast, the inside surface of elbow A must have been under condition similar to those pipes that were within the reach of orifices or flow nozzles as described in the preceding chapters. This estimation of the cause of the accident directly leads to the countermeasure for this case: do not use T junction but Y shaped branch pipe of carbon steel for the pure water flow at the dangerous temperature range of active/passive shift.

Another example is the case of heat recovery boiler for the waste gas of 155 °C. Serious wall thinning (3. 2 mm in 6. 5 years or 0. 5 mm y^{-1}) as well as the penetration of holes in the wall occurred at the roots of the heat conducting tube of carbon steel (Fig. (**7-21**) and (**7-22**)). The DO level of the boiler feed water had been maintained under 7 ppb, and the pH 8. 5 - 9. 5.

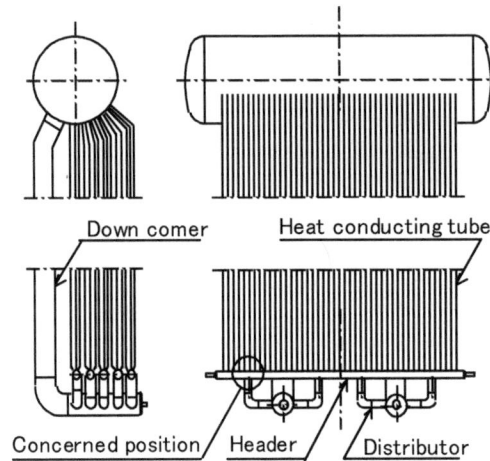

Fig. (7-21). Headers and heat conducting tubes in boiler.

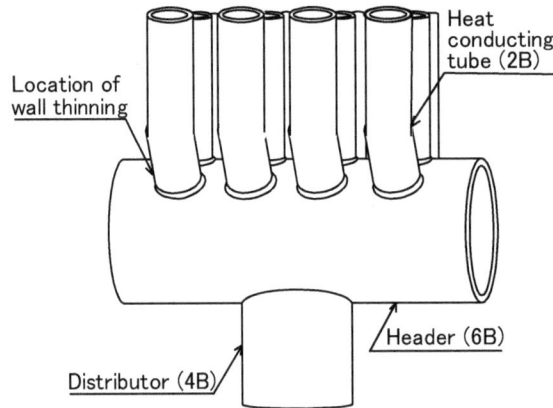

Fig. (7-22). Location where wall thinning occurred.

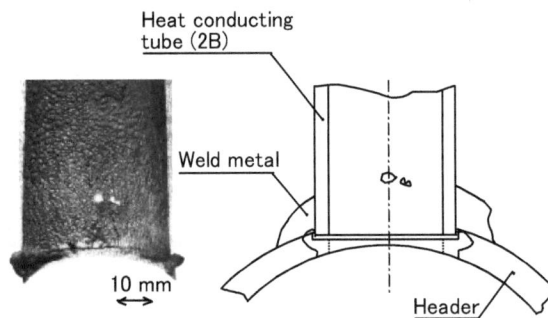

Fig. (7-23). Macrograph of inside wall surface of damaged tube near the tube-header joint.

In Fig. (**7-23**), the macrograph of the inside wall surface of the damaged tube near the tube-header joint as well as the sketch of the view are shown. The observation of these reveals two kinds of damage: the first is the comparatively uniform wall thinning which resulted in the opening of holes. The micrograph shown in Fig. (**7-24**) is of the wall surface at the downstream from the opened holes, which is seemingly the initial stage of the comparatively uniform wall thinning: a lot of scallop shaped indentions (1-3 mm in diameter) are arranged in the same direction which closely resembles to those in Fig. (**7-3**), which were assumed to be originated by the fixed vortexes under the deflecting jet (Figs (**7-5**) and (**7-6**)).

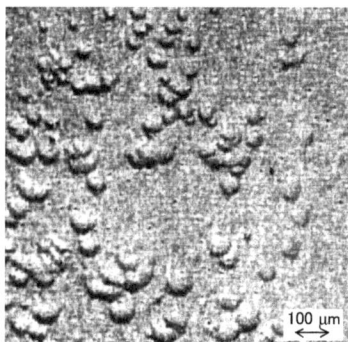

Fig. (7-24). Scallop shaped indentions on the inside wall surface at downstream from the pierced holes.

The second is the corrosion damage located in the pocket which was formed between the bottom end of the heat conducting tube and the seat hole in the header. Since it lies deep under the thick weld metal, it may take some more time to reach the surface. Nevertheless, the penetration with a rather high rate so far (0. 4 mm y^{-1}) should not be overlooked.

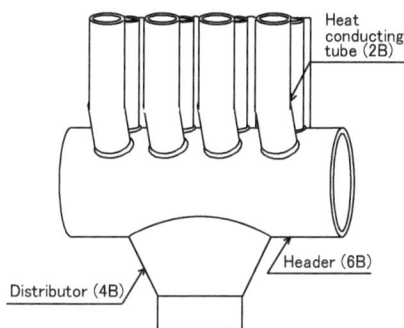

Fig. (7-25). Spreading of distributor exit in order to avoid the generation of dead water region.

The countermeasure proposed for preventing the wall thinning described above is avoiding the generation of dead water region by spreading the distributor exit in order to change the direction of water flow at the entrance of the heat conducting tube (Fig. (**7-25**)). In order to eliminate the pocket, the improvement in the welding joint which connects the tube with the header is proposed (Fig. (**7-26**)).

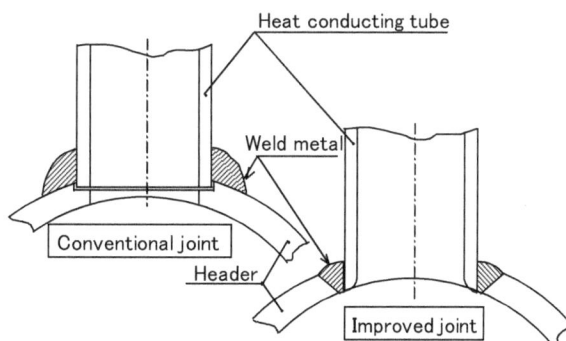

Fig. (7-26). Improve in tube-header welding joint.

Temperature gap across the critical temperature for active/passive shift of carbon steel inevitably produces a great difference in the anodic dissolution rates, which surely results in the formation of the macro-cell of corrosion, and accordingly the wall thinning of high rate. This mechanism is clearly and in detail illustrated with the corrosion rate *vs.* temperature diagram in Fig. (**6-23**). The typical examples of the case are given in Table **6-1**: in the case of Mihama nuclear power plant, the rupture occurred at the pipeline A which was equipped with the supporting rod. But, the pipeline B which paralleled A without supporting rod escaped from rupture. According to the heat transfer calculations, the temperature difference of about 10 °C was estimated between the pipe wall direct under the rod and the neighboring wall, with the assumption that the rod was not thermally insulated. This substantial temperature difference must have been the origin of the higher wall thinning rate.

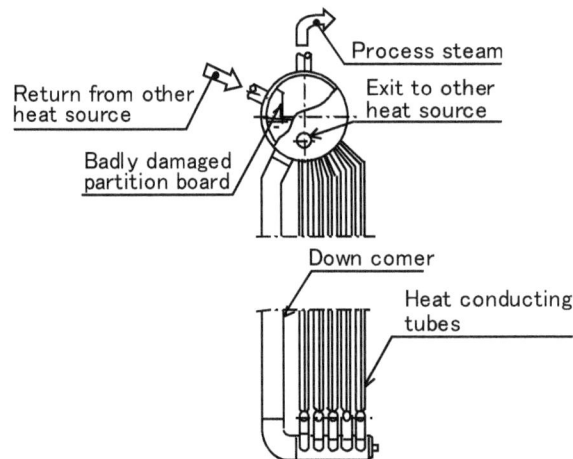

Fig. (7-27). Partition board in the drum of boiler.

In Fig. (**7-27**) the case of erosion-corrosion damage to a partition board is shown, where the temperature difference might be originated due to a careless process design. In the drum, the temperature of the water as well as the components including the partition board was assumedly higher than the critical temperature for active/passive shift of carbon steel. While, the temperature of the water flow which returned from the other heat source to the drum must be, from some reasons, lower than the temperature of the partition board. The temperature difference between the place where the water collides and the surrounding was assumed to be the origin of the active/passive type macro-cell of corrosion leading to the severe corrosion damage occurred on the board of carbon steel.

5.2. Elimination of Irregularities in Plant Operations

The accidental occurrence of erosion-corrosion presumably due to the irregularity in plant operation was displayed in Section 3. 6 of Chapter 6: it occurred on the carbon steel pipe in one of the three boilers of identical set-up under apparently identical operation conditions. The reason why the macro-cell of corrosion occurred only in the concerned boiler but not in other two was supposed as follows: the water flow stopping period in the intermittent operation at the concerned bypass pipeline must have been accidentally and irregularly held longer for the pipeline of the concerned boiler. The temperature difference thus brought about in the pipeline must have originated the active/passive type macro-cell of corrosion.

According to the experiences of engineers in the field, the incidents concerning erosion-corrosion tend frequent after the periodical shut down of the plant as well as the change of operating conditions. This is rationalized as follows: according past experiences, once the passivity is established on a metal surface, it would not collapse, even the environmental conditions change a little; nevertheless, the passivity may collapse in a reducing atmosphere which may be some times utilized for the prevention of general corrosion during the shutdown of plants for regular maintenance. If the operation were started after the shutdown with changes in environmental temperature, the surface might be impossible to return to passivity, and a macro-cell of the active/passive type may be formed, causing localized wall thinning at a higher rate.

In order to avoid such failure mentioned above, the corrosion states of inside wall of pipeline have to be monitored each time operation is resumed.

6. MONITORING AND SAFETY

Detection of the Inception and Activity of Macro-Cell: At the start up of plant operation, the inception of active/passive type macro-cell in the pipe wall surface of carbon steel may be detected through monitoring the potential of the pipe wall. Fig. (**7-28**) shows the potential variation with time for the jet-in-slit test specimen of carbon steel: the potential was stabilized soon at the pH 10, where the whole specimen surface became passive, but not at pH 9. 5 where passive and active coexisted.

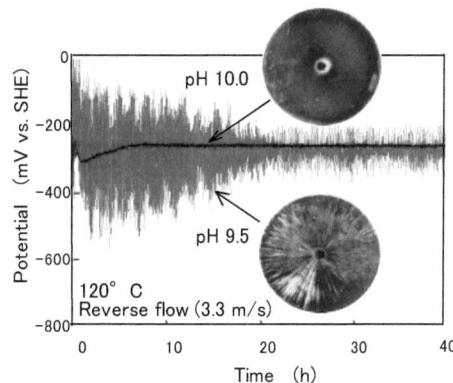

Fig. (7-28). Fluctuation of potential for jet-in-slit test specimen of carbon steel.

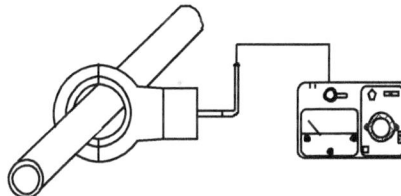

Fig. (7-29). Clip-on ammeter for monitoring the activity of macro-cell of corrosion.

The electrolytic corrosion in buried pipelines is sometimes monitored with the macro-cell current through a clip-on ammeter (Fig. (**7-29**)) or measuring the potential difference between the places apart a certain distance along the pipeline (Fig. (**7-30**)). These methods may be also utilized for detecting the activity of the macro-cell occurred in the pipelines.

Prevention of Pipe Rupture following the Concept of Leak before Break (LBB): Fig. (**7-31**) shows so-called "telltale hole" which functions to prevent rupture in the similar way that the safety valves work in pressurized vessels and pipelines:

A tiny hole, a few millimeters in diameter and in depth, board halfway in the wall of the pipe and open to the external surface. The pipe wall may be corroded from the inner surface, and the thickness may be decreased until it reaches the bottom of the hole, then the service fluid inside may blow out to give alarm.

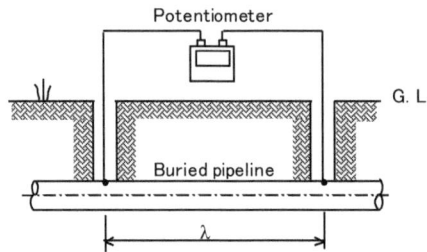

Fig. (7-30). Measuring the potential difference for detecting the macro-cell current.

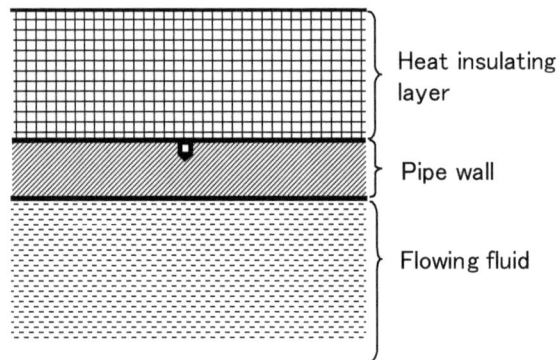

Fig. (7-31). Telltale hole to give alert for wall thinning.

7. INTEGRATED COUNTERMEASURE TO AVOID SERIOUS EROSION-CORROSION ACCIDENT IN BOILER

In order to avoid serious erosion-corrosion accidents in the boilers installed in the thermal and nuclear power plants as well as chemical plants, the following series of countermeasure should be taken.

1. Selection of construction materials for those boilers should be based on the results of jet-in-slit test which rather easily reproduces the localized corrosion of issue on materials under various conditions of water chemistry as well as water flow conditions in boilers.

2. In designing the piping plan, the importance should be put for a smooth flow without the rapid change at the flow velocity avoiding irregularities in fluid flow, e. g. the formation of stagnant watershed at downstream from flow nozzle and orifice or T junction should be avoided.

   A uniform temperature distribution over the piping system should be achieved avoiding not the hot spots but rather cold spots which used to occur near the mounting locations of pipe support rods and thin branch pipes.

3. The plant should be operated stably avoiding fluctuations in operation conditions such as the flow rate and the temperature of pipe wall.

4. The macro-cell current flow in the pipe wall should be monitored to detect the inception of macro-cell of corrosion, and its activity thereafter.

5. After rather longer operation period, the periodical shutdown maintenance should be conducted with the visual as well as the nondestructive inspections which are concentrated on

such highly suspected area as mentioned above: the cold spots as well as the places near bends and downstream from the flow meters where the stagnation of fluid flow apt to arise.

6. Mitigate the activity of macro-cell of corrosion with electrolytic protection technique, in particular, apply the anodic protection for carbon steel pipes suffered from active/passive type macro-cell corrosion, and eliminate the irregularities in pipe wall temperature distribution by completing the thermal insulation of the pipe.

7. The fail safe system following the concepts of safety valve or "leak before brake" should be introduced.

8. The top leader and the staffs as well as the operators should be encouraged to study fundamental corrosion engineering as well as the past and latest erosion-corrosion incidents in the field.

REFERENCES

[1]　Heitz E. Chemo-Mechanical Effects on Flow on Corrosion. Corrosion 1991; 47:135-145.

[2]　Matsumura M. The possibility for formation of macro-cell corrosion in a liquid with low electrical conductivity. Mater. Corros. 2011; 62: 449-453.

[3]　Fujita K. In: The 10th Course for Materials and Environments. Okayama: Chugoku-Shikoku Branch of JSCE. 2004; pp. 31-35.

[4]　Muramatsu T, Hibara H, Murakami S, Sudo K. Flows in T-junction Piping System (2nd Report, Numerical Analysis of Vortex Street Formed by Branch Pipe Flow). Bull. Jpn. Soc. Mech. Eng. 2004; 70: 2551-2558.

[5]　Kitashima N, Ichikawa K, Kinoshita K, Miyasaka M. Corrosion in Seawater Pumps and Its Prevention. Boshoku-Gijutsu (presently Zairyo-to-Kankyo) 1986; 35: 633-641.

[6]　Kastner W, Erve M, Henzel N, Stellwag B. Calculation code for erosion corrosion induced wall thinning in piping systems. Nucl. Eng. Des. 1990; 119: 431-438.

INDEX

A

activation polarization 84
all-volatile treatment 102, 138
anode 81
anodic current 81
anodic polarization curve 84
anodic protection 137, 148
average friction coefficient 42

B

banana vortex 79, 80
boundary layer 86, 87
breakaway velocity 11, 101, 120

C

cathode 81, 82
cathodic control 86
cathodic current 81
cathodic polarization curve 85
cathodic protection 137, 144
cavitation 3, 16
cavitation corrosion 5
cavitation erosion 16
cavitation intensity 29, 30
cavities 3, 5
cementite 124
characteristic depth 21, 22, 28, 47
combined water treatment 103
commercial pure iron 35, 50
concentration polarization 84, 85
corrosion cell 81
corrosion current density 86
corrosion potential 85
cracking process 32
critical impact velocity 40, 41, 42, 43, 44
current stopping potential 146
cut surface 35, 36
cutting process 32, 40

D

damage depth 21, 97
deformation process 32, 40
deflecting jet 140, 141
deposit attack 10, 89
dezincification 97, 141
dezincification proof brass 95
differential aeration-cell corrosion 88
differential flow-velocity corrosion 11, 89, 123
differential oxygen concentration-cell 88
displaced surface 35, 36

dynamic friction coefficient 41
dynamic pressure 73

E

electrode 81
electrolytic corrosion 145
equilibrium potential 84
erosion 5, 16
erosion corrosion 5
erosion·corrosion 10, 70, 101
erosion-corrosion 16, 94, 99, 102, 136
erosive wear 30
Evans diagram 85
extrapolation method 68

F

fail-safe 138
Faraday constant 82
ferrite 124
fish-scale pattern 135, 136
fixed vortex 78, 152
Flade potential 84
flow-accelerated corrosion (FAC) 14, 102
flow induced localized corrosion 102, 137
flow induced macro-cell orrosion 137
free jet 71, 72, 74
fountain jet apparatus 62

G

galvanic corrosion 87, 88
graphite 124
graphitization 11, 124
graphitization corrosion 124
graphitization layer 124, 125
grey cast iron 124
groove corrosion 114

H

half cell 81
hematite 81, 107
hill and valley 54, 72
horseshoe corrosion 10, 79, 80, 89

I

impingement attack 10, 89
impinging jet 71
impulsive pressure 3, 4, 5, 30
incubation period 9, 20
inhibitor 138, 139
inlet tube corrosion 10, 77, 89, 92
iron hydroxide 81

J

jet-in-slit 71, 72, 94
jet-in-slit apparatus 37, 102
jet-in-slit with reverseflow 77, 78

L

lepidocrocite 134
leak befor break 154
lip surface 35, 36
liquid impact erosion 46
liquid jet 4
localized corrosion 80, 81, 89
localized wall thinning 129, 131

M

macro-anode 83
macro-cathode 83
macro-cell current 83, 91
macro-cell current effect 91,131
macro-cell model 83
macro-cell of corrosion 11, 81
magnetite 13, 81, 107
micro-cell corrosion 81
micro-cell model 81
multi-cell 82
multi-cell model 82

N

neutral water treatment 103

O

oxygen cluster 82
oxygen diffusion limiting current density   85, 89

P

passivity 84, 130, 131
passivation 130, 131
pearlite 124
polarization 84
polarization curve 84, 85
polarization resistance 84

R

rain erosion 46
rapid test methodology 16
residue layer 125

S

scallop appearance 136, 139, 152
shear stress 73, 74
single electrode 81
skeleton corrosion 124

sliding surface 35, 36
slurry erosion 62
slurry erosion corrosion 31
stagnant watershed 129
static friction 41
static friction coefficient 41
static pressure 75
stationary specimen vibratory cavitation test facility 6, 7
strain hardening 22, 28
strain hardening index 27, 28, 30
stray-current electrolysis 144
stray electric current 137
submerged jet 71, 72, 76
surface area ratio effect 90, 92, 132
surface increment percentage 24, 28
synergistic effect 66

T

telltale hole 155
total pressure 75
turbulence 75, 76
turbulence corrosion 10
turbulent force 77

U

uniform corrosion 80, 81
uniform wall thinning 129, 131

V

vibrating specimen facility 6, 50
vibratory cavitation testing equipment 5
vortex rod 115
vortex-string 114, 115

W

water chemistry 137, 138
water tunnel 5, 6

Y

6/4 yellow brass 72, 76